Changes in the Landscape

Changes in the Landscape

Humans and Nature in Nineteenth-Century Latin America

Edited by
JENNIFER L. FRENCH

VANDERBILT UNIVERSITY PRESS
Nashville, Tennessee

Library of Congress Cataloging-in-Publication Data

Names: French, Jennifer, editor.
Title: Changes in the landscape : humans and nature in nineteenth century
    Latin America / edited by Jennifer L. French.
Description: Nashville, Tennessee : Vanderbilt University Press, 2024. |
    Includes bibliographical references and index.
Identifiers: LCCN 2024025227 (print) | LCCN 2024025228 (ebook) | ISBN
    9780826507457 (paperback) | ISBN 9780826507464 (hardcover) | ISBN
    9780826507471 (epub) | ISBN 9780826507488 (pdf)
Subjects: LCSH: Latin American literature--19th century--History and
    criticism. | Human ecology in literature. | LCGFT: Literary criticism. |
    Essays.
Classification: LCC PN849.L29 C47 2024 (print) | LCC PN849.L29 (ebook) |
    DDC 809.4/998--dc23/eng/20240809
LC record available at https://lccn.loc.gov/2024025227
LC ebook record available at https://lccn.loc.gov/2024025228

Excerpt from *Entranced Earth: Art, Extractivism, and the End of Landscape*, by Jens
Andermann. English translation copyright © 2023 by Northwestern University.
Published 2023 by Northwestern University Press. A previous version was published in Spanish in 2018 under the title *Tierras en trance: Arte y naturaleza después
del paisaje*. Copyright © 2018 by Ediciones Metales Pesados. All rights reserved.

Front cover image: Juan Manuel Blanes, *Gaucho en el campo*, Colección Viviane y
Enrique Manhard, Photo by Nicolás Vidal and Lorena Larriestra.

To Mary Louise Pratt, a force of nature.

Contents

A Rapidly Changing Landscape

Ecocriticism as an Approach to the Cultural Production of Nineteenth-Century Latin America

JENNIFER L. FRENCH

The aim of this book is to bring the methodologies and commitments of ecocriticism to bear on the study of Latin American literature and cultural production of the long nineteenth century. The essays collected here engage with nature—that most ubiquitous trope, from Bolívar (and before) to Martí (and after)—in ways that move beyond the deconstruction of political mythologies toward a more thorough understanding of the complex and multiform transformation of relations among humans and nonhuman nature that occurred during a period when capitalism's draw on resources was intensifying very rapidly in the region and around the world. In this introductory essay, I will define some key concepts and terms, starting with ecocriticism, contextualize the chapters that comprise *Changes in the Landscape: Humans and Nature in Nineteenth-Century Latin America*, and introduce them in the order that they appear.

What is ecocriticism, and what are its particular methodologies and commitments? As Cheryll Glotfelty wrote in her now-classic introduction to *The Ecocritical Reader* (1996), "all ecological criticism shares the fundamental premise that human culture is connected to the physical world, affecting and affected by it," and that "literature does not float above the material world in some aesthetic ether, but, rather, plays a part in an immensely complex global system, in which energy, matter, and *ideas* interact."[1] Critics have responded to the question of precisely how ecocriticism engages with materiality in a multitude of ways—including within the essays collected here— but another constant of ecocriticism is that it "by definition, inhabits a space on the periphery of the more traditional disciplines within the humanities, drawing on the work of colleagues in the natural and social sciences; it is (also by definition) an activist approach to the study of literature and culture."[2] As ecocriticism has become increasingly mainstream, disciplinary boundaries have at once shifted and become more porous, so that what are often referred to as the environmental humanities now encompass and/or dialogue with a broad range of theoretical frameworks and methodological approaches, including "green" Marxism, posthumanism, the critique of biopolitics, the ontological turn in anthropology, plant studies/the vegetal turn, animal studies, the new materialisms, the energy humanities, and so on and so forth.[3] All of these currents (and more) flow through the chapters of *Changes in the Landscape: Humans and Nature in Nineteenth-Century Latin America*.

History's Natures

As Raymond Williams pointed out many decades ago, "We need and are perhaps beginning to find different ideas, different feelings, if we are to know nature as varied and variable nature, as the changing conditions of a human world."[4] Williams was referring to the philosophical separation between Man and Nature that constitutes the foundation of the modern Western episteme, an idea that emerged in the colonial matrix of the sixteenth century Atlantic world and was famously codified in the writings of Francis Bacon and René Descartes. The critique of the modern Western episteme and the pursuit of alternatives have since become central projects of the environmental humanities. Indeed, the need that Williams identified in his essay "Ideas of Nature" has grown all the more urgent since then, as the consequences, for planetary life as a whole, of modalities of thinking and practice—economic and scientific, but also political and aesthetic—that permit the bracketing off of some segments of reality, be they species, populations, locales or phenomena, have become overwhelming and terrify-ingly real. That said, the cultural discourse produced by nineteenth-century Latin American creoles is not the first place many people would look for a critique of the modern Western episteme, much less alternatives. In fact, as Adrian Taylor Kane has recently noted, Latin American ecocriticism found its first foothold in the period immediately *after* the nineteenth century, that is, in the corpus of regionalist narrative that began to appear in the early twentieth, when the economic dislocations of the European war made the ecological degradation created by the export boom a topic of concern in many countries.[5] With some important exceptions, most scholarship on the nineteenth century has emphasized creole elites' complicity in the overar-ching historical processes of a period that opens with the movements for political and economic independence from the Iberian powers, only to close with brutal wars of conquest and burned-out internal frontiers caused by the rapacious overharvesting of rubber, yerba, bananas, timber, tannin, and other primary products.[6]

This trajectory makes the nineteenth century no less interesting or sig-nificant for Latin American ecocriticism. *Changes in the Landscape* covers a temporal arc spanning from the immediate aftermath of the Spanish-American Wars of Independence (1810–1826) to the early twentieth cen-tury (1925). This includes the period of global history Eric Hobsbawm called the Age of Capital, when the economic changes associated with the indus-trial revolution outpaced and overwhelmed the political transformations, based on Enlightenment principles of liberty, equality, and the impartiality of the law, that had been promised by the French Revolution and the Inde-pendence movements in the Americas.[7] In Latin America it was an especially violent time, because, as Hobsbawm puts it, "the very process of global cap-italist expansion . . . multiplied tensions within the overseas world, the am-bitions of the industrial world, and direct and indirect conflicts arising out of it."[8] In the new republics, where creole elites had sought the end of pen-insular control in order to escape the strictures of the Spanish trade monop-oly, the desired economic opportunities were slow to materialize while, as in the Empire of Brazil, cheap manufactures poured in from overseas, espe-cially British textiles that drowned artisanal production in much of the re-gion. Throughout much of Spanish America, Independence was followed by civil wars between conservatives whose power base was rural and who favored extending the traditional privileges of the Catholic Church and the more urban liberal factions that advocated secularization, constitutional government, and a broad agenda of modernization.[9] With the gradual consolidation of Liberal hegemony, foreign investment and loans financed infrastructure to facilitate the transport of primary products from the interior to overseas markets. By the late nineteenth century, with Liberal regimes firmly entrenched throughout much of the region, governments waged war against populations and territories they considered resistant to the march of progress and the expansion of capitalism into hinterlands that had largely avoided subjugation by the Spanish and Portuguese governments.

As environmental historian Guillermo Castro Herrera writes, economic stagnation—or even "ajuste a la baja," (downward adjustment)—in the early national period meant that in many parts of Spanish America the drain on nonhuman nature was equal to or even somewhat less than it had been prior to Independence.[10] That would change dramatically after 1850: with their urban power base, Eurocentric orientation, and modernizing agendas, the ruling elites in the second half of the nineteenth century eagerly oversaw dramatic transformations. Ports were dredged, steamships bought, and railroad tracks laid in an effort to "conquer time and space" through the burning of fossil fuels.[11] The new machinery would serve to deliver Latin American produce and other raw materials—beef, bananas, timber, tannin, rubber, coffee, copper, wheat, etc.—to northeastern Europe for the purpose of feeding a newly industrialized proletariat and the factories in which they labored.[12] This process culminated in the ecocidal export boom of 1870–1930. Historically and ecologically, capitalism developed as Jason W. Moore has recently described it: "co-produced by human and extra-human natures in the web of life; and cohered by a 'law of value' that is a 'law' of Cheap Nature. As the core of this law is the ongoing, radically expansive, and relentlessly innovative quest to turn the work/energy of the biosphere into capital (value-in-motion)."[13] Throughout much of Latin America, regions that were previously located outside the realm of capitalization were rapidly and violently incorporated into the new export economy. In many areas, campesinos were forced off lands their families had traditionally farmed, creating the labor power required to harvest timber, yerba and other forest resources, which rapidly led to soil depletion and droughts. Often the agricultural lands were themselves the main prize, and campesinos were forced onto ecologically fragile parcels so that elites could dedicate richer soils—while they lasted—to intensive monocropping.[14] Environmental historian Shawn William Miller describes a process of "civilization as blitzkrieg, one in which humans no longer tied themselves to a place called home but wedded themselves with utmost fidelity to profit. The result was a spreading cancer, ravaging everything at its perimeter and leaving a black, dead core characterized by deforestation, erosion, and ghost towns."[15] The banana industry, with its intensive use of toxic pesticides to control as much as possible the advance of fungus, lay waste to vast stretches of territory in Central and South America.[16] Most deadly of all was the wild rubber boom in the Amazon basin, in which tens of thousands of workers, many of them Indigenous Amazonians enslaved by rubber barons, were murdered in the early years of the twentieth century.

As the essays gathered in *Changes in the Landscape* demonstrate, the archive of nineteenth century Latin America offers an abundance of discursive crystallizations of the dramatic transition from one form of historical nature—Iberian colonialism—to another, the laissez-faire capitalism of the export boom. I am borrowing here Moore's use of "historical nature," which he in turn takes from Marx and Engels in *The German Ideology*.[17] Historical natures are nature, which includes both human and nonhuman being, as specifically inflected at specific historical moments, such that economic and biophysical processes are linked to concepts, aesthetics, and scientific practices; revolutions and other significant transformations tend to occur in all of these areas more or less simultaneously.[18] Crystallizations of historical nature are not necessarily straightforward and simple, since the representations themselves are part of the technics that at once mystify nature and make it known to us in any given period and place. As Carlos Fonseca has recently written, the landscape form that dominates nineteenth-century aesthetics is well understood as working to conceal the violence and incoherence of the nation-state, which attains a sheen of legitimacy through the harmonious forms contained in the image.[19] That we may still be capable of seeing an image such as Ferdinand Bellermann's *En el trapiche* as an essentially "natural" scene suggests the lasting influence of the landscape form, which was considered essential to the cultivation of taste and sensibility in eighteenth century Europe and spread to the Americas soon after.[20] In scenes such as Bellermann's depiction of a Venezuelan sugar mill, the realities of agricultural production all but disappear against the aestheticized treatment of lush tropical foliage and the dramatic changes of elevation; the eye hardly perceives the group of dark-skinned laborers who work a patch of cleared earth below the horses. Contributing to the effect is the conventional use of perspective in Western art, which invites the viewer to survey the scene from a singularly privileged position. "The visible world is arranged for the spectator as the universe was once thought to be arranged for God," in John Berger's apt description.[21]

An Archive of the Virtual

Scholars have been engaged in analysis and critique of the landscape form in nineteenth-century Latin American narrative and visual arts for many years now. We build on this work, especially Mary Louise Pratt's groundbreaking *Imperial Eyes: Travel Writing and Transculturation*, which first focused critical attention on the alignments among images and accounts produced by scientific travelers, creole elites' projects of national self-definition, and the economic exploitation of human and nonhuman nature in the post-Independence period.[22] Nevertheless, there are significant distinctions, methodological as well as theoretical, between the essays collected here and earlier scholarship. As befits ecocritical approaches, our methodologies are more broadly interdisciplinary than is usually the case in nineteenth-century studies, a field in which the influence of the linguistic turn has been especially pronounced.[23] More precisely, the essays in *Changes in the Landscape* engage with scientific writing as both a discourse of power and a source of potentially significant, even revelatory information about human and nonhuman nature. Whereas Gabriela Nouzeilles's introduction to her 2002 collection *La naturaleza en disputa: Retóricas del cuerpo y el paisaje en América Latina* begins with an epigraph from Derrida—"No hay naturaleza, sólo sus efectos" (There is no nature, only its effects)—our investigations depart from the premise of acute imbalance in the shifting assemblage of human and nonhuman being, matter and life, signs, affects and images that philosopher Gilles Deleuze refers to as "the real."[24] For us, it is well past the time for the study of literature and culture to excuse itself from engaging with materiality on theoretical or methodological grounds, even as we remain vigilant regarding the hierarchical relations embedded in discourses of nature and the ways that these have been (and are still) wielded in the Americas. We are working in the productive tensions among ecocritical, poststructuralist, and decolonial approaches.

It is obvious to many today that the problem—by which I mean the inability of the countries of the global north and the US in particular to mount anything like an effective response to the unfolding ecological catastrophe—is not the fault of our literary criticism or even primarily our science,

FIGURE 0.1. Ferdinand Konrad Bellermann (1814–1889), *En el trapiche* (*At the Sugar Mill*), c. 1868–70. Oil on canvas, 144.8 × 182.9 cm (57 × 72inches). Colección Patricia Phelps de Cisneros. Photograph by Gregg Stanger.

but rather a result of deep-seated dysfunction in the economy and political system. But there are nevertheless choices before us, including choices by which we may refuse to cede the response to the technological solutions devised by even the greenest capitalism, with its inherently inequitable distribution of environmental risk and security. To the extent that we can, and must, advocate for more radical responses than either political parties or conventional environmental organizations are likely to propose, the ways that we describe, consider, analyze, and imagine the past as well as the future matters very much in the framing of issues before us.[25]

In recent years, one of the key debates that has occupied scholars in the environmental humanities is the designation of the name "Anthropocene" to indicate the advent of a geologic age that can be distinguished from the previous one by perceptible evidence, in the stratigraphic record, of human activity beginning in the late eighteenth century, a date that coincides with the invention of the internal combustion engine by a Scottish engineer named James Watt. As has been repeatedly observed, the term "Anthropocene" projects an unjustifiably homogeneous concept of humanity—as if we were all equally responsible for the dangers of a warming planet—and invites renewed fantasies about the techno-heroics of Modern Man.[26] Less frequently addressed is the critical issue of periodization: the privileging of 1784 as the onset of the new era in planetary history isolates the burning of fossil fuels as the singular cause of the ongoing ecological crisis. This may bear a measure of truth in the case of climate change, but as Moore points out, allowing Stoermer and Crutzen's intervention to stand as the predominant framing of complex processes of ecosystem decline that involve not only temperatures rising because of the increase of atmospheric carbon but also deforestation, the Pacific trash vortex, algae bloom, the toxification of groundwater and soils, mass extinction, and more, sets up the search for "solutions" in terms of technological fixes, particularly the development of alternative fuels that would allow the global economy to churn on without fundamentally altering basic patterns of resource use. The alternative model Moore and others are advocating focuses instead on a much longer history of "humanity's modern relation with the rest of nature," a seminal event of which is the expansion of capitalism in the Atlantic world of the sixteenth century.[27]

Viewed through Moore's alternative periodization, the history of the Capitalocene, the essays gathered here offer snapshots of the transition between one mode of historical nature and another, from Iberian colonialism to the laissez-faire liberal economics of the export boom. It was not, of course, a clean break: one of the profound lessons of Latin American history is that Independence was not followed by decolonization, but rather as José Martí wrote, "La colonia continuó viviendo en la república" (The colony lived on in the republic). The stories told and retold by our contributors trace an anthropodecentric history of the Latin American nineteenth century, one in which the shifting relationships among the more-than-human occupy the center rather than the margins of the page. Our research also draws attention to the imaginaries that were operative, the technics that were available, and the choices that were made by relatively privileged people, such as the Mexican feminist Laura Méndez, in Catalina Rodríguez's account of her trade journal, the *Revista Hispano-Americana*, at a time when the structures and rules that previously governed their interchange with nature had been dramatically altered. *Changes in the Landscape* helps us to better understand the transition from colonial regimes to liberal extractivism by highlighting some of the cognitive resources that were available to imagine, facilitate, and enact new norms and expectations with regard to the relations between human and nonhuman life, be it the life of wildflowers, waterfalls, birds, or Cuba's Ciénaga de Zapata. Some of these essays draw out the contradictions within what would subsequently become dominant discourses, such as Gisela Heffes' analysis of writings by Domingo F. Sarmiento, José Hernández, and Juan Bautista Alberdi. Others, including interpretations of William Henry Hudson by Jens Anderman and Lesley Wylie, instead examine the intellectual and material alternatives to mainstream Western science and the capitalist logic of extraction that Latin American creoles began to imagine and pursue.

"The past," writes Elizabeth Grosz, "is always already contained in the present, not as its cause or its pattern but, rather, as its latency, its virtuality, its potential for being otherwise. This is why the question of history remains a volatile one, not simply tied to getting the facts of the past sorted out and agreed on."[28] In the ongoing crisis of the present, the sense of disaster's inevitability is paralyzing. For that reason, the chapters collected in *Changes in the Landscape* represent a new hybrid in that they seek to *create* as well as to critique: to draw out into the present and future moments in which the archive offers up the gift of something new and worthy of perpetuation. These moments of virtuality represent what Grosz calls "*conceivable futures.*"[29] The scholars whose work is represented here amplify these moments by inviting unexpected interlocutors to the conversation, as when Vanesa Miseres sets Soledad Acosta de Samper and other *criolla* botanists in conversation with Luce Irigaray or when Ronald Briggs introduces Simón Rodríguez to present-day hydrographers in the Andean region. As Briggs remarks, Rodríguez's "critique of the republic's failure to break out of the economic and social forms of Spanish colonialism does not become a limiting principle for his vision of the future." It is an idea we should all keep in mind as we consider our capacity for solidarity and experimentation.

Summary of Chapters

The collection opens with "Canals, Dams, and Colonized Landscapes: Simón Rodríguez vs. the Vincocaya Project (Arequipa, 1830)," in which Ronald Briggs returns to the Venezuelan educational reformer—and Simón Bolívar's much-admired tutor—to reconsider Rodríguez's critique of an engineering project that was proposed to transform the landscape surrounding the Peruvian city. The project in question, a plan to dam the Chili

River and divert its waters for agricultural use, was a joint public/private venture that as Rodríguez points out, would have enabled Arequipa's elite to increase the value of their own lands while potentially desiccating the surrounding area. As Briggs demonstrates, Rodríguez—whose counter-proposal involves the construction of a canal that would draw on local knowledge and traditional land-use practices—understands "infrastructure and landscape as a space for staging the future." Briggs reads Rodríguez's critique of the Vincocaya project in dialogue with discourse on dams and water use from twentieth and twenty-first century sources as well as Marx's contemporary concerns about the ecological impacts of large-scale agriculture, suggesting that the future Rodríguez has in mind is based on a progressive if arguably paternalistic vision of engineering for equitable, sustainable communities.

In Chapter 2, "Forests of Sound: Listening, Affect, and Matter in Humboldt and Hudson," Jens Andermann turns to writings by two foundational figures in the history of Western environmentalism, the German geographer Alexander von Humboldt and the Argentine naturalist William Henry Hudson. Andermann compares Humboldt's 1849 essay "The Nocturnal Life of Animals in the Jungle," an immersive soundscape that represents a significant departure from the visual ekphrastic rhetorical mode for which Humboldt is better known, with Hudson's accounts of birdsong in *Birds of La Plata*, *Idle Days in Patagonia*, and other works. Engaging Mary Louise Pratt's influential reading of Humboldt as well as Timothy Morton's critique of Western environmentalism's persistent perception of "nature" as something external to the human self, Andermann teases out and analyzes the ways these writers rely on sound as "embodied, sensory experience in the field" in their construction of scientific argument. In this ultimately Deleuzian interpretation of Humboldt and Hudson, the shift from visual to aural registers allows for the recognition—albeit more persuasively, Andermann notes, in Hudson's writings than in Humboldt's—of what he calls "a dissident geo- and anatomopolitics of sound," including the liminal experience of becoming-animal.

Chapter 3 is Gisela Heffes' powerful essay, "Archives of Extinction: Unproductive Bodies and Human/Nonhuman Expendability in the Argentine Desert." Heffes examines the geographical concept at the center of Argentina's national identity, the *desierto*. Conventionally translated as "desert," the term, as Heffes explains, reduces the complex and diversely populated ecosystems of the Argentine interior to *terra nulius*, a vacant space waiting to be colonized after the defeat of Dictator Juan Manuel de Rosas in 1852. Examining the works of three key figures of nineteenth century Argentine literature—Domingo F. Sarmiento, José Hernández, and Juan Bautista Alberdi—Heffes teases out their contrasting attitudes regarding bodies and landscapes deemed unproductive according to the modernizing, capitalist logic of the State. As the landscape was rapidly transformed through the construction of railroads, the installation of barbed wire, the transition from ranching to agriculture, and (after 1879) the genocidal *Conquista del Desierto*, the representations of gauchos and other life-forms in their writings offer what Heffes calls an "archive of extinction." The chapter concludes with a discussion of the contemporary novel *Las aventuras de la china Iron*, published by Gabriela Cabezón Cámara in 2017, and its rewriting of the pampa as a space of connectivity teaming with human and nonhuman life.

In Chapter 4, "Cuba's Ciénaga de Zapata: Despoiled Landscapes and Biodiversity Conservation in the Long Nineteenth Century," Lizabeth Paravisini-Gebert traces the ecocultural history of Cuba's 6000-square kilometer Ciénaga de Zapata, a place of spectacular natural beauty that encompasses large areas of salt marshes, estuaries, floodplains, mangrove forests, and Caribbean coastal zones of unsurpassed biodiversity. The chapter opens with a speculative reflection on José Martí's childhood experience in the swamp, where his father was stationed by the Spanish Crown in order to stifle—within limits—the illegal traffic in enslaved Africans. In addition to Martí's earliest surviving letter, Paravisini-Gebert considers accounts of the swamp's transformation in the late nineteenth and early twentieth centuries, including writings by environmental engineer J. A. Cosculluela, novelist Luis Felipe Rodríguez, and Pura de Armas, whose family eeked out a precarious living by producing charcoal and availing themselves of the turtles, crocodiles, and other protein sources offered by the swamp. In the late nineteenth and early twentieth centuries the swamp was threatened with being completely drained and many of its native species were vulnerable to extirpation or extinction. The chapter ends with a brief coda about the success of the Cuban Revolution's conservation efforts, which led to the revitalization of the Ciénaga, the protection of its endemic species, and its concomitant designation as a UNESCO World Heritage site in the year 2000.

In Chapter 5, "Botanical Beings: On Women, Flowers, and Plants in Nineteenth-Century Latin America," Vanesa Miseres explores the varying ways that upper-class women engaged with the field of botany in the latter half of the nineteenth century, including scientific study, collecting, needlework, games, drawing, and painting. She examines in particular the albums that women created and circulated among their social acquaintances, where the "language of flowers" expressed coded sentiments. Miseres focuses on works by two Colombian women, María Josefa Acevedo (*El oráculo de las flores i el de las frutas*, 1857) and Soledad Acosta de Samper (*Conversaciones y lecturas familiares*, 1896). As Miseres demonstrates, the Linnaean system, which classifies plants according to their reproductive parts, reinforced traditional expectations regarding heterosexual gender roles. However, the same botanical pastimes that served certain disciplinary functions also provided upper-class women with a "rare opportunity" to venture into the masculine world of science and offered a welcome outlet for those seeking ways to establish and exercise agency over their sexual and reproductive bodies. Miseres's reading concludes with a productive dialogue between her nineteenth-century interlocutors and feminist philosopher Luce Irigaray.

In Chapter 6, "Hydraulic Energy, Nation, and Modernization in Nineteenth-Century Mexico," painting, lithography, and photography guide Jorge Quintana Navarrete's examination of two key transitions in Mexico's historical use of hydraulic energy. The first represents Mexico's initial wave of industrialization in the 1860s, when rivers were harnessed to power factories in the Valley of Mexico amid rising concerns about the ecological and economic effects of deforestation. The second transition occurs in the 1890s, when Mexico's coal and oil deposits were deemed insufficient to power the expansion of electricity for domestic and industrial use and the Porfirio Díaz regime constructed the Necaxa powerplant amid a landscape populated by Indigenous communities and dominated by the towering Necaxa/Huachinango Falls. Drawing on David Nye's concept of the "technological sublime" and very recent developments in the emerging field of the energy humanities, Navarrete's reading of visual texts and surrounding discourses carefully reconstructs the nineteenth-century history of President Andrés Manuel López Obrador's recent commitment to modernize Mexico's more than sixty state-owned hydroelectric plants.

In Chapter 7, "'That Mysterious *Something*': Nature, Mystery, and Animism in W. H. Hudson's Early Writings," Lesley Wylie explores the autodidact Argentine naturalist's persistent mistrust of Western scientific rationality and the alternative ways of knowing he proposes. This deeply researched chapter situates Hudson within his complex geographic and cultural contexts, including his formative years in Argentina and Uruguay and his writing career in late nineteenth-century England. Wylie draws out

Hudson's temperamental alignment with Romanticism and his proclivity for legitimating traditional, local knowledge over the claims of more formally trained men, be it in the form of "fairie stories" from the UK or animal ethology as interpreted by his Patagonian interlocutors. In addition to demonstrating Hudson's privileging of his own (and others') embodied experiences of nonhuman nature with examples from the rich store of anecdotes in Hudson's publications, Wylie engages with Hudson's writings from the standpoint of posthumanist theory and contemporary debates regarding the rights of nonhuman animals. She builds on scholarship by Sara Castro-Klarén by delving into Hudson's own definition and experience of "animism," which she likens to the phenomenon of Amazonian perspectivism described by anthropologist Eduardo Viveiros de Castro.

Chapter 8 is Aarti S. Madan's "Estanislao S. Zeballos, or Nineteenth-Century Argentina's Environmental Unconscious." As Madan explains, Zeballos was both an architect of the extermination of the originary peoples inhabiting Argentina's southern territories—the euphemistically named *Conquista del Desierto* (1878–85)—and one of the founding figures in Argentina's conservation movement. Given Zeballos's own communication with Theodore Roosevelt and other North American contemporaries, Madan sets Zeballos's work into an inter-American theoretical framework by drawing on writings from ecocritic Lawrence Buell. She distills Zeballos's extensive body of work—which includes a trilogy of nostalgic Indianist novels as well as geographical treatises and the military manual *La conquista de quince mil leguas* (1878)—to essential passages that document the changing perceptions of a bad faith actor who nevertheless becomes a prominent voice in decrying the causal relationship between changing land-use patterns and the increased risk of drought and flooding in the conquered territory. This unlikely advocate, Madan notes, offers extensive comments on climate change's differential impacts on subsistence farmers as opposed to the large-scale landholders, including British capitalists and Argentine veterans of the *Conquista del Desierto*, who controlled vast parcels of terrain in the southern territories.

In Chapter 9, "*La Revista Hispano-Americana* (1895–1896): Laura Méndez's Extractive Pedagogy," Catalina Rodríguez examines fiction and nonfiction works by the feminist writer, a prominent figure in Porfirian Mexico. The chapter opens with Méndez's story about a Mexican-American landowner who is swindled out of his property by his Anglo tenant ("La curva" or "The Curve," 1908), which Rodríguez reads as a cautionary tale about the hazards of privileging sentimental attachment over more modern and entrepreneurial modes of relating to the land in the aftermath of the US Intervention in Mexico (1846–1848) and the annexation of California. *The Revista Hispano-Americana* addressed similar concerns, Rodríguez writes, and "endorsed an instrumental relationship between humans and nonhuman nature in an attempt to address the disparities between Latin and Anglo-Americans." Rodríguez examines the bilingual publication in terms of the messages it sent its dual audiences: for Anglo readers, the *Revista* offers "a catalogue" of attractive business opportunities in Latin America, while for Latin American elites it provided an education in the wealth to be gained from successfully capitalizing the region's abundant natural resources. The chapter closes with a discussion, based on writings by ecofeminist Val Plumwood, of the links between Méndez's work as a feminist educator and the marketing of feminized landscapes in her short-lived periodical.

Chapter 10 is Jennifer L. French's "Memories of a Darwinian: Anarchism and Animality in the Literary *Crónicas* of Rafael Barrett." Building on insights from the critic Leandro Delgado, French's reading of the Hispano-Paraguayan anarchist teases out Barrett's idiosyncratic Darwinism from his exposé of labor abuses in the yerba mate industry (*Lo que son los yerbales*, 1906) and the literary-scientific *crónicas* he published in the newspapers of Asunción, Montevideo, and Buenos Aires in the early years of the twentieth century. French maintains that Barrett's aestheticized discourse represents a "Darwinism without Darwin"—a Darwinism from which Darwin himself is largely absent—that may be productively interpreted in dialogue with modern and contemporary biophilosophy from writers including Henri Bergson, Elizabeth Grosz, Donna Haraway, and Gilles Deleuze. In Barrett's vibrant and eclectic writings, the understanding that human and nonhuman animals are separated by degrees of difference rather than stark ontological divides prompts intermittent passages in which a spontaneous intuition of amorphous connection to other life-forms arises, a phenomenon Grosz calls "becoming undone." These moments enable Barrett to engage in a trans-specific affective community across vast extensions of time and space, a move that echoes his critique of capitalist exploitation across human and nonhuman life-forms.

In the final chapter, Emmanuel Velayos examines "Graffiti as Earthly Inscriptions: Human Acts and Geological Forces in Euclides da Cunha's *Os sertões* (1902)." This meticulously reasoned essay opens with a discussion of da Cunha's 1902 article "Fazedores do desertos" (The Desert-Makers) that explains how farming practices, swidden in particular, had brought severe droughts to the formerly lush tropical region of southern Brazil. The article, in which da Cunha refers to humans as a "nefarious geological agent," sets the stage for *Os sertões*, the book-length chronicle of the Brazilian army's war against the renegade religious community living under the leadership of Antonio Conselheiro in the city of Canudos. As Velayos explains, da Cunha's famously eclectic book—which contains extensive sections on the physical environment and its human inhabitants—effects a "two-sided translation," by which geological forces are represented in terms of human agency while human dynamics are simultaneously converted into geological ones. Velayos follows this elaborate rhetorical operation through da Cunha's account of a wall covered in layers of graffiti, through which Brazilian soldiers express their misery and contempt. For Velayos, the writing and rewriting of this and other scenes surrounding the Canudos massacre enable da Cunha to articulate his skepticism regarding the uses to which positivist science was put by the Brazilian intelligentsia. Importantly, the chapter dialogues with Dipesh Chakrabarty and other contemporary scholars regarding the entanglement of human and nonhuman forces implicit in the concept of the Anthropocene.

NOTES

1. Cheryll Glotfelty, *The Ecocriticism Reader*, xviii–xix.
2. French and Heffes, *The Latin American Ecocultural Reader*, 5.
3. Foundational texts on these topics can be found in the references following this introduction.
4. Williams, *Problems in Materialism and Culture*, 85.
5. Adrian Kane, preface to *The Natural World in Latin America Literatures*; also see Jennifer L. French, *Nature, Neocolonialism, and the Spanish-American Regionalist Writers*.
6. See, for example, Madan, *Lines of Geography in Latin American Literature*; Martínez-Pinzón, *Una cultura de invernadero*; Uriarte, *The Desertmakers*; Beckman, *Capital Fictions*; and the "Commodities" section of DiGiovanni and Uriarte, *Latin American Literature in Transition, 1870–1930*. For the exceptions, see, for example, Lesley Wylie's reading of the death of a ceiba tree in Andrés Bello's "Silva a la agricultura de la zona tórrida" in *The Poetics of Plants in Spanish American Literature*, 17–52; and Steven F. White's study of Rubén Darío in *Arando el aire: La ecología en la poesía y la música de Nicaragua*. See also chapter 4, "Nature and the Nation-States," in *The Latin American Ecocultural Reader*, 83–124.
7. Hobsbawm, *The Age of Capital, 1848–1875*. The classic study of this period in Latin American history is found in Halperín Donghi, *Contemporary History of Latin America*, 74–208.
8. Hobsbawm, *The Age of Capital*, 78.
9. See Martínez-Pinzón and Uriarte's introductory essay in *Entre el humo y la niebla*, 22–23.
10. Castro Herrera, *Naturaleza y sociedad*, 149–224.
11. Miller, *An Environmental History of Latin America*, 139.

12. Miller, *An Environmental History of Latin America*, 106.
13. Moore, *Capitalism in the Web of Life*, 14.
14. Castro Herrera, *Naturaleza y sociedad*, 203–24.
15. Miller, *An Environmental History of Latin America*, 131.
16. See Martínez-Pinzón, "Bananas and Plantains" in *Latin America in Transition*, 59–73. The volume edited by DiGiovanni and Uriarte contains detailed studies of literature and other cultural discourses related to several of the export boom's most prominent commodities: rubber, guano and nitrates, coffee, plantains and bananas, sugar, and yerba. See DiGiovanni and Uriarte, *Latin America in Transition*, 13–102.
17. Moore, *Capitalism in the Web of Life*, 11–12.
18. Carolyn Merchant's classic book *Ecological Revolutions* illustrates the concept of historical nature through an ecofeminist study of colonial New England.
19. Fonseca, *The Literature of Catastrophe*, 15–16.
20. Luz Aurora Pimentel writes, "For Latin America, language—be it Portuguese or Spanish—was an imported reality; so too were the literary and pictorial schemes used in the representation of the real, not to speak of the political, social, intellectual and domestic networks that organized life in the colonized lands." From "The Representation of Nature in Nineteenth-Century Narrative and Iconography," in *Literary Cultures of Latin America*, 62–109; see also Maderuelo, *El paisaje*; and Marrero, ed., *Transatlantic Landscapes*.
21. Berger, Blomberg, Fox, Dibb, and Hollis, *Ways of Seeing*, 16.
22. Pratt, *Imperial Eyes*.
23. In addition to Pratt's *Imperial Eyes*, major works of scholarship in the area of nineteenth-century studies were published and/or translated during the apex of poststructuralism: Doris Sommer's *Foundational Fictions*; Ángel Rama, *The Lettered City*; and Roberto González Echevarría, *Myth and Archive*.
24. Nouzeilles, *La naturaleza en disputa*, 11; Colebrook, *Deleuze*, 66.
25. See Stengers, *In Catastrophic Times*.
26. See Haraway, *Staying with the Trouble*, especially 30–57; and Pratt, *Planetary Longings*, 117–24.
27. Moore, *Capitalism in the Web of Life*, 173.
28. Grosz, "Histories of a Feminist Future," 1020. I am grateful to Mary Louise Pratt for introducing me to Grosz's work.
29. Grosz, "Histories of a Feminist Future," 1020.

BIBLIOGRAPHY

Ades, Dawn. *Art in Latin America*. Yale University Press, 1989.

Beckman, Ericka. *Capital Fictions: The Literature of Latin America's Export Age*. University of Minnesota Press, 2013.

Bennett, Jane. *Vibrant Matter: A Political Ecology of Things*. Duke University Press, 2010.

Berger, John, Sven Blomberg, Chris Fox, Michael Dibb, and Richard Hollis. *Ways of Seeing*. BBC and Penguin, 1972.

Burkett, Paul. *Marx and Nature*. St. Martin's, 1999.

Castro Herrera, Guillermo. *Naturaleza y sociedad en la historia de América Latina*. CELA, 1996.

Colebrook, Claire. *Deleuze*. Routledge, 2002.

Coole, Diana, and Samantha Frost, eds. *New Materialisms: Ontology, Agency, Politics*. Duke University Press, 2010.

de la Cadena, Marisol. "Indigenous Cosmopolitics in the Andes." *Cultural Anthropology* 25, no. 2 (April 2010): 334–70.

DiGiovanni, Fernando, and Javier Uriarte. *Latin American Literature in Transition, 1870–1930*. Cambridge University Press, 2022.

Fonseca, Carlos. *The Literature of Catastrophe: Nature, Disaster, and Revolution in Latin America*. Bloomsbury Academic, 2020.

Fornoff, Carolyn, and Gisela Heffes. *Pushing Past the Human in Latin American Cinema*. SUNY Press, 2021.

Foster, John Bellamy. *Marx's Ecology*. Monthly Review Press, 2000.

French, Jennifer L. *Nature, Neocolonialism, and the Spanish-American Regionalist Writers*. University Press of New England, 2005.

French, Jennifer, and Gisela Heffes, eds. *The Latin American Ecocultural Reader*. Northwestern University Press, 2021.

Gagliano, Monica, John C. Ryan, and Patricia Vieira, eds. *The Language of Plants: Science, Philosophy, Literature*. University of Minnesota Press, 2017.

Giorgi, Gabriel. *Formas communes: Animalidad, cultura, biopolítica*. Eterna Cadencia, 2014.

Glotfelty, Cheryll. "Introduction: Literary Studies in an Age of Environmental Crisis." In *The Ecocritical Reader: Landmarks in Literary Ecology*, xv–xxxvii. Edited by Glotfelty and Harold Fromm. University of Georgia Press, 1996.

González Echevarría, Roberto. *Myth and Archive: A Theory of Latin American Narrative*. Cambridge University Press, 1990.

Grosz, Elizabeth. *Becoming Undone: Darwinian Reflections on Life, Politics, and Art*. Duke University Press, 2011.

Grosz, Elizabeth. "Histories of a Feminist Future." *Signs* 25, no. 4 (Summer 2000): 1017–21.

Halperín Donghi, Tulio. *The Contemporary History of Latin America*. Translated by John Charles Chasteen. Duke University Press, 1993.

Haraway, Donna J. *Staying With the Trouble: Making Kin in the Chthulucene*. Duke University Press, 2016.

Hobsbawm, Eric. *The Age of Capital, 1845–1876*. Vintage, 1996.

Kane, Adrian Taylor. Preface to *The Natural World in Latin American Literatures: Ecocritical Essays on Twentieth Century Writings*, edited by Adrian Taylor Kane, 1–8. McFarland, 2010.

Madan, Aarti S. *Lines of Geography in Latin American Literature*. Palgrave, 2017.

Maderuelo, Javier. *El paisaje: Génesis de un concepto*. Abada Editores, 2005.

Marder, Michael. *Plant-Thinking: A Philosophy of Vegetal Life*. Columbia University Press, 2013.

Marrero, José Manuel, ed. *Transatlantic Landscapes: Environmental Awareness, Literature, and the Arts*. Instituto Franklin, Universidad de Alcalá, 2016.

Martínez Pinzón, Felipe. *Una cultura de invernadero: Trópico y civilización en Colombia (1808–1928)*. Iberoamericana/Verveurt, 2016.

Martínez Pinzón, Felipe, and Javier Uriarte. *Entre el humo y la niebla: Guerra y cultura en América latina*. University of Pittsburgh, 2017.

McCance, Dawn. *Critical Animal Studies: An Introduction*. SUNY Press, 2021.

Merchant, Carolyn. *Ecological Revolutions: Nature, Gender, and Science in New England*, 2nd edition. University of North Carolina Press, 2010.

Miller, Shawn William. *An Environmental History of Latin America*. Cambridge University Press, 2007.

Moore, Jason W. *Capitalism in the Web of Life*. Verso, 2015.

Nouzeilles, Gabriela. Introduction to *La naturaleza en disputa: Retóricas del cuerpo y del paisaje en América Latina*. Paidós, 2002.

Pimentel, Luz Aurora. "The Representation of Nature in Nineteenth-Century Narrative and Iconography." In *Literary Cultures of Latin America: A Comparative History*, vol. 2, edited by Mario J. Valdés and Djelal Kadir, 156–72. Oxford University Press, 2004.

Pratt, Mary Louise. *Imperial Eyes: Travel Writing and Transculturation*, 2nd ed. Routledge, 2009.

Pratt, Mary Louise. *Planetary Longings*. Duke University Press, 2022.

Rama, Ángel. *The Lettered City*. Translated by John Charles Chasteen. Duke University Press, 1996.

Sommer, Doris. *Foundational Fictions: The National Romances of Latin America*. University of California Press, 1993.

Stengers, Isabelle. *In Catastrophic Times: Resisting the Coming Barbarism*. Open Humanities Press/Meson, 2015.

Szeman, Imre, and Dominic Boyer, eds. *Energy Humanities: An Anthology*. John Hopkins University Press, 2017.

Uriarte, Javier. *The Desertmakers: Travel, War and the State in Latin America*. Routledge, 2020.

Viveiros de Castro, Eduardo. "Cosmological Deixis and Amerindian Perspectivism." *Journal of the Royal Anthropological Institute* 4, no. 3 (September 1998): 469–88.

Waldau, Paul. *Animal Studies: An Introduction*. Oxford University Press, 2013.

White, Steven F. *Arando el aire: La ecología en la poesía y la música de Nicaragua*. 400 Elefantes, 2011.

Williams, Raymond. "Ideas of Nature." In *Problems in Materialism and Culture*, 67–85. Verso, 1980.

Wolfe, Cary. *What Is Posthumanism?* University of Minnesota Press, 2009.

Wylie, Lesley. *The Poetics of Plants in Spanish-American Literature*. Pittsburgh University Press, 2020.

CHAPTER I

Canals, Dams, and Colonized Landscapes

Simón Rodríguez versus the Vincocaya Project (Arequipa, 1830)

RONALD BRIGGS

In 1830 the Venezuelan educational reformer Simón Rodríguez published in Arequipa a pamphlet entitled "Observaciones sobre el terreno de Vincocaya con respecto á la empresa de desviar el curso natural de sus aguas y conducirlos por el rio Zumbai al de Arequipa" (Observations on the terrain of Vincocaya with respect to the project of diverting the natural course of its waters and directing them to the river of Arequipa by way of the Sumbay River).[1] Framing his essay as a response to Mariano E. Rivero and Clemente Althaus's "Reconocimiento de la obra de Vincocaya" (Recognition of the Vincocaya work) written during the same year, Rodríguez argued that the Vincocaya Project, which proposed to dam the Colca River and divert the water south to the Chili River where it could be used for irrigation, was a flawed design for a purpose that could be better accomplished by a canal.[2] He concluded with a rhetorical question very much in keeping with the style of his better-known educational texts: "¡¿Quién creería que una ACEQUIA / diese motivo para escribir tanto!?" (Who would believe that an IRRIGATION DITCH / would be a reason for writing so much?!)[3]

The question asks his readers to reflect on the extensive sociopolitical musings that accompany his technical arguments in favor of the canal. By the end of the text, he has made an argument not only for how a portion of the Andean highlands might be irrigated but also for the future of Arequipa and Spanish American independence writ large.

I have written elsewhere about Rodríguez's scientific writings on the Vincocaya Project and the earthquake and tsunami that destroyed Concepción, Chile, in 1835, comparing Rodríguez's perspective with those of Alexander von Humboldt and José María Blanco White, with a focus on harmony and the social and political implications of physical landscape.[4] Here I propose to put Rodríguez's text in dialogue with the social and scientific discourse on irrigation both in his own time and in our own present moment, defined by climate change and increased episodes of drought and scarcity on the one hand, and intense storms and flooding, on the other. I will be arguing that Rodríguez's critique of Vincocaya and his counter proposal suggest pathways for rethinking colonized space both in terms of immediate benefits and reforms as well as from the perspective of future patterns of development. While he could not foresee the present moment, when climate change would literally erase year-round snow atop El Misti that served as Arequipa's literal and spiritual reservoir during the nineteenth century, Rodríguez's project attempts to view infrastructure and landscape as a space for staging the future, with all of the contradictions and problems that this task entailed.[5]

Rodríguez's concern for the human labor (and laborers) that makes irrigation possible and the agricultural labor that necessarily follows causes him to consider relationships between labor and capital in both the short term context of construction and the imagined future to follow. Resisting the idea of monoculture crop development on large plots of land, he makes an argument for a varied landscape, what we might call an artificial natural balance, as a sustainable and profitable alternative. Rodríguez acknowledges the dynamic in which Indigenous people and Indigenous labor were othered by European and Criollo notions of progress, and his pedagogical focus on the future allows some space for thinking outside of racialized castes, even as it remains circumscribed by the distance and authority he assumes in the role of teacher. His view of development as something that can be incorporated into the idea of nature suggests new ways of imagining the economic and social future outside the binary of urban versus rural or development versus nature.

Arequipa: "Dominance" and "Limitation"

In Rodríguez's context, independence and revolution are very much part of the historical backdrop against which Arequipa's landscape is to be considered. The Vincocaya Project itself had been launched formally in 1826, barely a year after the Battle of Ayacucho (December 1824) had secured Peru's independence from Spain. It took shape as a public/private venture by a group of 84 wealthy landowners in Arequipa.[6] The Vincocaya Project promised to significantly raise the value of the land that would suddenly become irrigated, and, as one observer has put it, "y de paso, mantener el estatus y prestigio social de algunas importantes familias" (and, in passing, to maintain the status and social prestige of some important families).[7] It could also be classified as an example of municipal corruption. Alfonso W. Quiroz has chronicled the self-dealing inherent in the relationships between investors and government officials and John Frederick Wibel has recounted that "Arequipeños sarcastically described the peninsular Pedro Murga, leading colonial merchant and Caylloma miner and the original manager of the Vincocaya Project, as the 'miner of coined silver of Vincocaya.'"[8] The "Acta de asociación del Estado y particulares" (Act of association of the state and individuals) for the Vincocaya Project was filed by Arequipa notary Pedro de Luque in March 1826. It detailed a significant collaboration between the government and the project's investors, promising that the government would provide assistance in general and specific terms: "El Gobierno se compromete a remover todos los obstáculos que se opongan a la empresa por interés privado o común, y a expedir las órdenes más eficaces para proporcionar los operarios, Maestros, y cuántos brazos sean necesarios para la Obra, de los lugares que se le pidan" (The Government commits to removing all of the obstacles opposed to the enterprise by private or public interests, and to expedite the necessary orders to proportion the machinists, supervisors, and as many workers as are needed for the job, from the places where they are requested). These instructions make it clear the question of labor was on the minds of the project's planners, and that part of the value of their investment depended on the government's ability to provide skilled and unskilled workers for the project.[9] Neither the original Vincocaya Project nor Rodríguez's alternative was ever completed, and a look at nineteenth-century and early twentieth-century sources reveals a fascinating continuation of the debate as engineering reports championed the possibility of a dam or agreed with Rodríguez about the dam's difficulties, often citing precedents from Europe and the United States and disputing technical points such as the risk of collapse from different designs and the degree of evaporation to be expected in a high-altitude, low latitude reservoir.[10] The relevance of this last concern, evaporation, is evidenced by a recent op-ed in the *New York Times* by Adam Lustgarten detailing a one year loss of "a million acre feet of water" to evaporation in reservoirs in the western US.

In 1971, nearly 150 years after the first iteration of the Vincocaya Project, the left-leaning military dictatorship of Juan Velasco Alvarado would launch the Majes-Siguas project to dam the Colca River in a different location and divert its waters for irrigation, a project that would blossom into "a grand hydraulic system consisting of seven dams, rivers and canal connections."[11] The Velasco government's framing emphasized the socio-economic consequences of the project, which purported to create new farmland to be distributed to small farmers and to help lead Peru "out of poverty and feudalism into an era of equality and progress."[12] Having entered its second phase and at best caught between its stated purpose and the realities of "an agro-energy project oriented towards high-tech agribusiness and hydroelectric production," today the Majes-Siguas project demonstrates the wide-ranging implications of landscape-altering infrastructure.[13]

Traditionally seen as a second city and rival to the Peruvian capital of Lima, Arequipa has been called "una oasis entre el desierto y la montaña" (an oasis between the desert and the mountains), and identified, regionally, by "fervent regionalism" and as "a center of royalist sentiments" and "resistance to both imperial and national authorities" with strong historical and economic links to the Republic of Bolivia.[14] Situated in the valley formed by the Chili River and occupying an average altitude of 2328 m, Arequipa is a high altitude city compared to coastal Lima but still considerably lower than cities on the Peruvian and Bolivian altiplano such as Cusco (3399 m), La Paz (3640 m), or Puno (3827 m), and it sits below much higher mountain peaks such as the volcanoes known as El Misti (5821 m), Chachani (6057 m), and Pichu Pichu (5664 m), occupying something like a physical middle position between the Peruvian Andes and the desert coast. Its mild climate meant, as Sarah C. Chambers has put it, "that the main hindrance to agriculture in Arequipa was the scarcity of water," and the Vincocaya Project became the first of many post-independence attempts to remove this hindrance by transforming the region's landscape.[15]

It is important to point out that this project of landscape transformation in Arequipa predates both independence and the arrival of Spanish colonizers in the sixteenth century. Observers have noted the implicit irony of any

attempt to see irrigation as a simple modernization of the Peruvian landscape, given the extensive systems of irrigation that had long been in place
when Europeans first arrived.[16] These systems, as one historical commen-
tator has put it, "desaparecieron rápidamente por la incuria de los conquis-
tadores" (disappeared rapidly from the conquistadors' neglect).[17] And if the
nineteenth century's irrigation projects brought with them the prospect of
a new water-controlling bureaucracy, this too was "reappearance" rather
than new creation, a republican version of pre-colonial systems of politics
and power based on water and agriculture.[18]

It is also important to point out Arequipa's traditionally European orientation, a factor that differentiates it from Andean cities like Cusco, Puno,
and La Paz. It has been called "the most Spanish and nonindigenous of
Peruvian cities," a moniker that takes on particular significance given its
proximity to the Andes and the large Indigenous and mestizo populations
in the surrounding regions.[19] It is perhaps not a coincidence that the April
8, 1826, issue of *El Republicano* that announced the Vincocaya Project contained another article describing the repression of a revolt of enslaved people in Havana as a case in which "se pusieron los blancos reunidos en defensa por su propia seguridad general" (the whites united in defense of their
own general safety). Where the Havana article employs the term *blancos reunidos*, the description of the project celebrates the power of *hombres unidos*:
"Así en hombres unidos no se resisten los cerros ni los rios" (Thus before
men who are united, neither the hills nor the rivers resist). In both cases
the collective project is a triumph of a "we" over an obstacle strictly defined as Other. By increasing the value of lands largely owned by a group
of wealthy white Arequipeños, the Vincocaya Project promises an agricultural and economic "conquest" of the surrounding landscape—a transformation of land outside the city proper that will increase the relative wealth
and power of its wealthiest and most powerful inhabitants. This binary
vision that opposes "nature" to a conquering enterprise reflects the inner
geographic tension of a city defined, in paradoxical terms, by "el sentido
de dominación y el sentido de limitación" (the sense of domination and
the sense of limitation).[20]

Making and Unmaking Deserts

Irrigation, a product of "engineers and their dreams," offers the prospect
of "'let the desert bloom' utopia."[21] In the case of Arequipa the literal push

and pull between domination and limitation can be transcribed onto the struggle to transform desert into cropland. This process or narrative of "let the desert bloom" is not the only possible chronology, as evidenced by the nineteenth-century exchange of letters on declining crop yields between Karl Marx and Friedrich Engels, and also in the testimony of a local resident interviewed by Astrid Stensrud in the context of Majes Siguas, a farmer who worried that the newest addition to the project would direct water to new locations by removing it from the old ones: "we could be transforming Majes into a desert."[22]

Javier Uriarte has pointed out that the very etymology of deserts suggests a chronology that runs in reverse of "let the desert bloom": "the desert, in its very etymology, *was not always there*."[23] Citing Tacitus citing a Scottish chieftain's complaint about the Roman empire—"and when they make a desert they call it peace"—Uriarte argues that modernity is defined by creating deserts rather than transforming them into something else.[24] This formulation, fitted as it is to warfare, might seem particularly out of place when applied to irrigation. However, as Foster has pointed out, Marx's discussion of the early nineteenth-century crisis of soil fertility, a crisis that would occasion Peru's own guano boom several decades later, leads him to conclude that agriculture, left to its own devices, creates rather than erases deserts.[25]

Marx was responding in 1868 to his reading of works on soil depletion and crop yield by Justus von Liebig and Karl Fraas. Surveying the propensity of large-scale agriculture, especially monoculture, to deplete the soil and thus make its own continued cultivation impossible, Marx concludes, in a letter to Engels, that "the first effect of cultivation is useful, but finally devastating through deforestation, etc." before proposing a general principle based on his reading of Fraas, a principle whose rhetoric mirrors that of Tacitus: "The conclusion is that cultivation—when it proceeds in natural growth and is not *consciously controlled* (as a bourgeois he [Fraas] naturally does not reach this point)—leaves deserts behind it."[26] Nothing like the crisis of soil fertility produced by years of ceaseless agricultural development would be applicable to the landscapes around Arequipa that would be newly planted as a result of the Vincocaya Project or Rodríguez's alternative canal. Uriarte's and Marx's conceptions of the relationship between development and the desert are nonetheless useful here because of Rodríguez's insistence on contemplating the future landscape that irrigation will produce.

Desert is a slippery term, as Uriarte's analysis reveals, in part because it is frequently used as a description of landscape and climate on the one hand, and because its etymology is tied to the presence or absence of human communities on the other. One definition of desert-making would be the creation of places where people no longer live, and, to the degree that human communities depend on agriculture, the presence or absence of one of the term's constitutive elements stands in for the presence or absence of the other. Jennifer L. French has made a detailed comparison and contrast between the use of the word "desierto" in Spanish and "wilderness" in English, emphasizing the voices of critics of Spanish colonialism such as Montesinos and Las Casas for whom "'desierto' is not the prior condition of the Indies so much as the terrifying reality that European men have created in their unreasonable desire to exact surplus profits from the bodies of Native Americans and the land."[27] This moral variation on the term suggests an additional layer of meaning in Marx (and perhaps also in Tacitus): the desert as a moral ruin produced by the exploitation of land and people. In French's reading the development of "monocrop agriculture on a national scale" leads to an interlocking series of changes in landscape and social structures that produce "devastating results for workers and the environment."[28] Rodríguez's resistance to the original plan of the Vincocaya Project might usefully be characterized as a moral and political choice to produce new versions of social and economic community rather than to reify the old ones.

Landscape as Future: "The Economy of the fields"

Simón Rodríguez's best-known writings, *Sociedades Americanas en 1828* (published in numerous editions and reiterations throughout his lifetime), *Luces y virtudes sociales* (1842) and *Defensa de Bolívar* (1830), advance a series of polemical arguments about the need to finish the independence project through a social and economic transformation fueled by public education. Frequently compared to early socialists, especially Robert Owen and Henri de Saint Simon, Rodríguez wrote of the need to build social consciousness through education.[29] With a social and political project already in mind by 1830, Rodríguez is primed to consider the future of Arequipa as a question that far exceeds the scope of land value and crop yields. Jorge López Palma has argued that the canal plan carries with it "sus esperanzas de redención social en la agricultura" (his hopes for social redemption in agriculture) and reflects a balance between "una mecánica interna socializante" (an internal socializing mechanism) and "la finalidad de promover el desarrollo económico" (the goal of promoting economic development), while Camila Pulgar Machado has explained Rodríguez's vocabulary of landscape transformation as a metaphor for larger, region-wide social changes.[30]

If it is possible, as Paul Gelles has suggested, to say that "irrigation systems are texts to be read," then Rodríguez's critique of the dam and support for a canal contains several layers of meaning that are not hidden within the text but rather unfold as part of its construction.[31] Brian Larkin's oft-cited observation about infrastructure systems in general, that they "also operate on the level of fantasy and desire," applies here, too, and by Rodríguez's lights the hypothetical canal wins in any fair comparison with the hypothetical dam not only because of the implicit danger of collapse and catastrophic flood that dams create, but also because of the differences in future landscapes and social relationships that the dam and canal can be expected to foster.[32]

Rodríguez's scientific objections to the dam project include the risk of collapse and the more gradual problem of evaporation from the reservoir the dam would create, and he counts a social cost against the dam too. By creating a new lake it would flood land that has its own positive uses, putting underwater "una considerable extension de terreno, constantemente cubierto de buenos pastos—donde pueden caber 25 estancias ó asientos pastoriles, ocupados por muchas familias Indíjenas, cuidando de un gran número de ovejas" (a considerable quantity of land, constantly covered with fodder, occupied by many Indigenous families, caring for a large number of sheep).[33] He adds a parenthetical note intended as a preemptive barb against potential critics: "Esto sería exagerado para quien no entienda la Economía de los campos" (This would seem exaggerated to anyone who doesn't understand the Economy of the fields).[34] What Rodríguez is outlining here is more than support for "fields" against mines or, say, factories, but rather in-depth and qualitative knowledge about how these "fields" really function at a scale often ignored by engineers.

Rodríguez makes a similar argument when he suggests using Indigenous expertise to cultivate the land between an older, existing irrigation ditch and the canal he proposes to build. If the Vincocaya Project has in effect failed to take into account the land it will subtract from Peru's economy, Rodríguez is determined to consider also the possible gains it has neglected:

La compañia debe pensar en utilizar el terreno escarpado que quedará entre la Acequía y el Canal, porque tiene agua con qué regarlo, y porque á pesar de cuantas precauciones se toman contra la infiltración, el suelo inferior se humedecerá á expensas del canal. En Arequipa se conoce muy bien el arte de cultivar suelos quebrados, cortándolos en gradas.[35]

The company should think about using the steep terrain that will remain between the Irrigation Ditch and the Canal, because it has water with which to irrigate it, and because no matter how many precautions you take against infiltration, the lower soil will dampen at the canal's expense. In Arequipa they know very well the art of cultivating uneven ground by cutting it into terraces.

Here Rodríguez admits both the inevitable imperfection of his project—canals cannot be perfect, even if dams have to be perfect—and suggests that the local knowledge needed to convert this loss into a net gain is already available. As in the previous passage regarding the land displaced by the lake, scale matters. Rodríguez is willing to think of the landscape in small increments rather than strictly on the scale of large agricultural development.

Instead of large, single-crop farms he foresees an economy built on interlinking flora and fauna on a smaller scale: "el lino, el cáñamo, los bosques, los prados artificiales, y las crias de ganado mayor y menor" (flax, hemp, forests, artificial meadows, and major and minor livestock husbandry).[36] Blurring the lines between "nature" and cultivation, Rodríguez proposes that the waters of the Colca River, routed through a canal, be used to create a mixed landscape for a combination of different crops and livestock. The landscape would also include forest, which might normally be considered unproductive. From Rodríguez's perspective the forests that irrigation could produce will prove more valuable than fields of wheat:

Los bosques, á mas de mantener húmedas las tierras y fertilizadas con sus despojos (¡cosa que el Perú necesita tanto!) darian resinas, alquitran, madera de construccion, corcho (cultivando el alcornoque de esta especie), y leña (aprovechando de las podas, recortes, y muerte natural de los árboles).

Forests, in addition to keeping the land humid and fertilized with its remains (something that Peru so needs!) will provide resins, tar, lumber, cork (cultivating the oak of this species), and firewood (taking advantage of the prunings, trimmings and deadfall from the trees).[37]

In this vision the forest provides everything from additional materials for agriculture and construction to a form of climate control and thus serves as a countervalent force against desertification in both the strictly climatic sense of the word and its broader moral and social meaning.[38]

Having outlined the possible use of irrigation to create an artificial version of forest, something normally considered as a "natural" landscape feature, Rodríguez goes on to explain how irrigation can similarly produce artificial meadows that likewise fit into an integrated local economy:

En los prados se podrían mantener colmenares—el ganado serviria en los campos y abasteceria el mercado—Las necesidades irian indicando otros ramos de la industria, de los muchos que nacen de los productos inmediatas de la materia animal. Es de advertir que la temperatura del valle de Arequipa, favorecerá lo dicho y mucho mas.

In the meadows it would be possible to keep apiaries—the livestock would be of use in the fields and would provide for the market—Necessity would go on indicating other branches of industry, from among the many that are born from the immediate products of animal material. It is worth noting that the temperature of the valley of Arequipa would favor the above and much more.[39]

Again, without directly mentioning human communities, Rodríguez presents a vision that presupposes a variety of human activities sharing the same space along with a mixture of local and more distant markets. While this vision presupposes a significant transformation of the landscape around Arequipa, it also frames itself by way of its connections to current conditions. In a sentiment similar to that already noted in Chambers's analysis, Rodríguez suggests that water alone is the missing element.

To sum up the sensibility behind what Rodríguez expects irrigation to do for landscape, we might cite the summation he himself offers early on in his text—a statement of purpose that comes before the more precise descriptions of the canal and its expected result. Speaking in general of his project, he couches its objective in terms of the creation of "habitable" rather than just "arable" land: "para hacer HABITABLE en todas sus puntos, una region que parece estar, en gran parte, condenada por la naturaleza á ser eternamente desierto" (to make HABITABLE in all of its points, a region that seems to be, in large part, condemned by nature to be eternally desert).[40] Having translated "desierto" as "desert," it's worth pointing out that, as Uriarte suggests, "deserted" would be an equally accurate and indeed more etymologically precise translation. Rodríguez's project, while self-consciously rooted in local knowledge, is anything but an appeal to undeveloped land or a return to idyllic notions of original natural or social communities. Late in the essay he saves particular vitriol for the romanticized vision of nature invoked by the pastoral aesthetic—especially the association of rural life as a more carefree or leisurely alternative to life in the city.[41] For Rodríguez the countryside is as much a space of work and labor relations as the city, and any contemplated renewal of landscape must take into account the question of human labor both within the project itself and on a long term scale.

Landscape and People

The link between irrigation, landscape, and society would become a commonplace of discourse on modernization during the twentieth century. In the United States, the director of the Tennessee Valley Authority, which built a system of dams used for flood control and hydroelectric power, would argue that "everywhere what happens to the land, the forests, and the water determines what happens to the people."[42] Cristina Rivera Garza's account of a mid-twentieth-century irrigation and dam project carried out by the Cárdenas government in the Mexican state of Tamaulipas would make a similar point: quoting a government official's view that the project itself could be "material para una gran novela mexicana" (material for a great Mexican novel) perhaps comparable to Fyodor Gladkov's 1925 Soviet realist novel *Cement*, which chronicled "los primeros esfuerzos de organización industrial que hizo la Rusia soviética" (the first efforts of industrial organization carried out by the Soviet Union).[43]

For Rodríguez the differing social impacts of the Vincocaya Project's proposed dam and the canal he seeks to build matter as much or more as the technical differences in terms of landscape and risk. Along with being "un peligro más o menos inminente" (a more or less imminent danger) the dam concentrates capital as surely as it concentrates risk.[44] A canal, on the other hand, necessarily touches land belonging to many different owners thus distributing more widely the burdens of upkeep. In the case of a dam, he argues, "solo el dueño soporta los gastos de reparación" (the owner alone bears the costs of repair), and the question of repairs is a high stakes endeavor undertaken against the risk of catastrophic collapse.[45] The canal, on the other hand, produces a situation in which "cada vecino tiene un interés en conservarlo, y lo hace porque le cuesta poco" (every neighbor has an interest in maintaining it, and they do it because it costs them little).[46]

Rodríguez proposes not only the construction of a canal, but that he himself assume the role of "maestro de obras" (master builder) or the supervisor charged with the hands-on responsibility of "hacer entrar en el menos tiempo posible, el mayor número de acciones posible" (making the greatest possible number of actions fit into the least possible time). This role is a pedagogical one in Rodríguez's view and depends (in an echo of his better-known writings on educational reform) not only on the master builder's knowledge but also on the ability to communicate that knowledge to the workers: "no basta que el maestro lo sepa; si el obrero no entiende el trabajo NO LUCE" (it's not enough for the teacher to know it; if the worker does not understand the work WILL NOT SHINE).[47] Having established that the work on a dam must indeed "really shine" in order for the construction to be safe, Rodríguez goes on to imagine the difficulties of dam construction from the perspective not so much of the worker as of the supervisor in charge of the work:

> ¿¡Qué habilidad . . . qué talento . . . no necesitará tener el maestro que haga *lucir su trabajo* con Indios medio-salvajes, convertidos en albañiles por la virtud de una elección?! El Gobernador y el Alcalde de un lugar tuvieran poder para hacer salir al pobre indio de su choza, y encaminarlo á Vincocaya; pero nó para infundirle, de repente, el arte de trabajar.

> What ability, what talent, would the teacher need to have in order to *make the work shine* with half-savage Indians, converted into brick masons by means of a decision? The Governor or Mayor of a place might have the power to force the poor Indian to leave his shack, to direct him to Vincocaya; but not to infuse in him the art of working.[48]

Rodríguez's use of the racist trope "half-savage" to describe the region's Indigenous population is particularly jarring given the attention he appears to give to Indigenous communities and traditions in other parts of the essay, and it cannot easily be explained away. A page later he will effectively rephrase this objection in less clearly racialized terms, summing up his labor-based objection in terms of "lo uno, la casi imposibilidad de conseguir que lo *hagan bien*, albañiles poco o nada versados en obras de esta especie—y lo otro, porque los peones han de encarecer la obra, mucho mas de lo que se piensa" (for one thing, the near impossibility of getting brick masons to *do it well*, when they have little or no experience in this type of work—and the other thing, because the workers will add costs to the work, far beyond what one might think).[49] Here he frames his objection less in terms of anything being particularly wrong with the local and largely Indigenous workforce and more in terms of the difficulty of producing a European-style masonry dam in a place with little history or tradition of constructing that particular kind of irrigation system.

It is possible to read in these lines a thinly veiled critique of the new republican order as nothing more than poorly disguised renewal of the colonial arrangements that relied on forced labor. Silvia Rivera Cusicanqui has analyzed the consequences of the internal contradictions of a republican arrangement that through various forms of enslavement and inequality habitually undermined its stated egalitarian principles.[50] When Rodríguez points out that governors and mayors can force a group of Indigenous workers to relocate themselves to Vincocaya to work on the dam but cannot instantly create a tradition of European craft, he is, in effect, associating the investors in the original Vincocaya Project with the hubris of colonialism. By this reading we might say that Rodríguez is offering an architectural or engineering version of the concept of "ideas out of place" in the Latin American context that would be coined and theorized by Roberto Schwarz in the twentieth century.[51] He also presents the project as a pedagogical challenge and himself as the authority who will make the workers succeed because of his unusual ability to teach.

This trope of the American public (in the case of Andean Peru a largely Indigenous public) as a problem to be solved through education is very much in keeping with the full body of Rodríguez's work and thought, as well as with that of other Criollo independence leaders such as Bolívar. In a 1825 letter to his former student and current patron, Rodríguez had said, "El pueblo es tonto en todas partes; sólo Ud. quiere que no lo sea en América y tiene razón" (The public is foolish everywhere; only you want it not to be so in America, and you are right).[52] Considered this way, the objectification of the figure of the Indigenous public represents a local adaptation to a general principle—that publics everywhere are imperfect and incomplete without the sort of education Rodríguez can offer. The notion of social change based on educational reform, either through schools or the training of workers, carries with it a certain distance between the authority who offers the education and the students. While Rodríguez does not repeat the colonialist move that Marisol de la Cadena has described in the thinking of John Locke—that of designating Indigenous peoples as too close to "nature" on an imagined binary with humanity—Rodríguez's pedagogical focus does function as a form of what Cadena calls "hegemonic politics," which "tells its subjects what they can bring into politics and what should be left to scientists, engineers, magicians, priests, or healers," if not altogether abandoned.[53]

This question of authority is also relevant to the issue of irrigation. Molle has used the term "secular priesthoods" to refer to the power wielded by those officials empowered with control of water distribution in both pre-Columbian times and in the projects undertaken in the nineteenth and twentieth centuries.[54] In the context of the Majes-Siguas Project, Paerregaard and Baez Ullberg speak of "hydrosocial community" to illuminate "the intersection between infrastructure projects and community formation," while elsewhere Ullberg has employed a variation—"hydrocracy"— to delineate the situation in which engineers and designers "force water users to adapt to their projects."[55] Hydrosocial communities and hydrocracy shift the focus from the project's construction to its long-term social effects and could thus be applied to Rodríguez's vision of a mixed landscape and a mixed economy driven more by networks of internal consumption than by the export of cash crops. This concern for social effects in the future is also evident in Rodríguez's suggestion of the appointment of water regulators who would be placed in charge of community distribution. These officials would be "personas decentes que entiendan en la parte esencial" (decent people who understand the essential parts), and their control would help avoid "discordia" (discord) within the community.[56] Rodríguez's canal project thus serves as a platform on which to articulate a progressive and to some degree paternalistic vision of social and economic reform as well as a critique of the Vincocaya Project as a republican endeavor likely to reify the social and economic order of Spanish colonialism. If it manages to sidestep the problem that Cant has identified with Majes Siguas—a tendency to move forward without considering "existing forms of production"—it runs the risk of imposing its own hegemonic vision on the local population.[57]

Conclusion

Hydrocracy and hydrosocial communities hearken back to Larkin's already cited connection between infrastructure and desire. If the planning and execution of infrastructure systems harness the desires of investors, planners, designers, and engineers to influence the public in terms of material reality and imagination, they also function as part of a feedback loop in which projects influence public desires and imagination that at least in theory could come to influence the future projects of investors, planners, designers, and engineers. Late in his text on Vincocaya, Rodríguez makes it clear that the canal-based irrigation system he is proposing should be taken as a prototype, a kind of small-scale pilot for the plurality of development ideas he expects to follow: "Abran los Arequipeños los ojos sobre su empresa. Muchos proyectos útiles están esperando para realizarse, el suceso de las aguas de Vincocaya y el de las de Tacna—el Departamento de Arequipa, por su suelo y por el jenio de sus habitantes está destinado á ser la Cataluña del Bajo Perú" (Let the Arequipans open their eyes to the enterprise. Many useful projects are waiting, in order to be realized, the outcome of the waters of Vincocaya and of Tacna—the Department of Arequipa, for its soil and for the genius of its inhabitants, is destined to be the Cataluña of Lower Peru).[58]

This declaration bookends something said early on in the text, when Rodríguez begs his reader's pardon, "Perdone la sociedad presente esta digresión, en favor de la futura" (Let the present society pardon this digression in favor of the future one), and suggests that the Vincocaya Project's greatest value would only be realized in the future: "despertaré ideas, y llamará la atención de los Peruanos sobre un género de industria, que debe ser, *por largo tiempo*, el objeto de sus especulaciones y el sujeto de sus tareas" (it will awaken ideas, and call the attention of the Peruvians to that category of industry that should be, *for a long time to come*, the object of its speculations and the subject of its business).[59] In both of these passages the tone is strikingly optimistic, a counterpoint to his pessimism about the technical difficulties of the dam itself. The elegant invocation of the future to justify a digression in the present encapsulates in a smooth rhetorical form the long-term sense of perspective Rodríguez wishes to add to the discussion of irrigation. His critique of the republic's failure to break out of the economic and social forms of Spanish colonialism does not become a limiting principle for his vision of the future.

Olevarría Araya has characterized Rodríguez's pluralistic vision of Spanish American societies as an antidote for Enlightenment-inspired intellectualism and its tendency to hold at a distance the public for whom it pretends to speak.[60] Like education, the development of infrastructure creates a point of contact and friction between intellectuals and the public. Stensrud has noted, in the contemporary context of Majes Siguas, how amid the vagaries of agricultural markets, many farmers express "a profound feeling of having no control over the future" despite their access to a dependable supply of water.[61] Rodríguez's text does not solve these problems, though the open-ended gesture toward unspecified future projects that might be inspired by Vincocaya presents, in its plurality, something like an acknowledgment of the limits of his own vision. He also demonstrates, at a very early stage of political independence, the degree to which it was already possible to think beyond binaries separating economic development from wilderness preservation. Rodríguez's instinct is to fight the desert with water, forests, pastures, and communities, rather than any single cash crop.

NOTES

1. Some of the research for this article was completed while I was a Fulbright Scholar in Peru in 2022. I am grateful to the Institute for Latin American Studies (ILAS) for funds that enabled me to travel from Lima to Arequipa and to Sarah Chambers, Thomas Love, and Mario Rommel Arce Espinoza for advice on archives in Arequipa. Special thanks to Christian Castelo Meza at the Archivo Regional de Arequipa for his invaluable assistance: he directed me to the correct notary and even the correct pages for the Vincocaya-related papers, which I would never have found without his assistance.
2. Rodríguez, *Obras completas*, 415–70.
3. Rivero y Ustáriz and Althaus, *Colección de memorias científicas*, 268–74.
4. Rodríguez, *Obras completas*, 470.
5. Briggs, *Tropes of Enlightenment*, 58–88.
6. Andersen, "Of Volcanoes, Saints, Trash, and Frogs," 31–45. Andersen uses the phrase "darkening peaks" to refer the visible specter of climate change in the mountain skyline over Arequipa (41) in an essay that also points out the necessary use of dams and reservoirs to serve as a water reserve in the era of diminished snowmelt (22–32).
7. Barriga, *Memorias para la historia de Arequipa*, 341.
8. Condori, "Economía y empresa," 204. Condori also notes that the 1826 launching of the Vincocaya Project also came on the heels of "la primera crisis del capitalismo" (the first crisis of capitalism) provoked by failed British investment in Latin America (188).
9. Quiroz, *Corrupt Circles*, 102; Wibel, "The Evolution of a Regional Community," 362.
10. "Acta de asociación del Estado y particulares," 320. I have modernized the spelling and accentuation of the original.
11. Tamayo et al., "Aguas del Chili," 553–58; Eduardo L. de Romaña, "Informe del ingeniero Don Romaña," 558–56; Hurd, "Estudiar para aumentar," 7–43. Tamayo and Hurd sup-

ported the basic Vincocaya Plan, suggesting that the original, never completed dam be broken up and used as fill (36). For his part, Romaña cites a recent dam collapse in California as evidence in support of Rodríguez's warnings about the dangers of a dam in comparison to a canal (563).

12. Andersen, "Of Volcanoes, Saints, Trash, and Frogs," 31–32.
13. Stensrud, "Dreams of Growth," 574. Stensrud is citing the rhetoric of the Velasco government.
14. Ullberg, *Water Alternatives*, 509.
15. Belaunde, *Memorias*, 10; Chambers, *From Subjects to Citizens*, 36; Wibel, "The Evolution of a Regional Community," 9, 3, 324. Chambers likewise refers to Arequipa as "a royalist stronghold to the end" (36).
16. Chambers, *From Subjects to Citizens*, 39.
17. Regal, *Los trabajos hidráulicos*. Regal notes that the region of Arequipa contained "un sistema de canales prehispánicos" (a system of pre-Hispanic canals, 141). Other references to the general loss of pre-Hispanic irrigation during the colonial period can be found in Regal (14, 49) and also in Paul Gelles, *Water and Power in Highland Peru*.
18. Leguía y Martínez, *Historia de Arequipa*, 93; Bruce Manheim and Guillermo Salas Carreño have traced the Quechua use of separate words to refer to "the substance of water" and "the flowing water of irrigation," in their essay "Wak'as: Entifications of the Andean Sacred," in *The Archaeology of Wak'as* (48).
19. Molle, Mollinga, and Webster, *Water Alternatives*, 329.
20. Love, *The Independent Republic of Arequipa*, xi.
21. Belaunde, *Memorias*, 11.
22. Stensrud, "Dreams," 573; Molle, *Water Alternatives*, 330.
23. Stensrud, "Dreams," 571.
24. Uriarte, *The Desertmakers*, 1.
25. Uriarte, *The Desertmakers*, 1.
26. Foster, *Marx's Ecology*, 150–51, 156–57.
27. Marx and Engels, "Marx to Engels," 559.
28. French, "Voices in the Wilderness," 159.
29. French, "Voices in the Wilderness," 162.
30. For more on Rodríguez's social and political significance see Jorge López Palma, *Simón Rodríguez: Utopía y socialismo*; and Leon Rozitchner, *Filosofía y emancipación*. For more on Rodríguez's life and educational project, see Alfonso Rumazo González, *Simón Rodríguez, maestro de América*; and Fabio Lozano y Lozano, *El Maestro del Libertador*.
31. López Palma, *Simón Rodríguez*, 113, 109; Pulgar Machado, *La materia y el individuo*, 136.
32. Gelles, *Water and Power in Highland Peru*, 7.
33. Larkin, *Annual Review of Anthropology*, 327–43.
34. Rodríguez, *Simón Rodríguez* 1:438.
35. Rodríguez, *Simón Rodríguez* 1:438.
36. Rodríguez, *Simón Rodríguez* 1:463.
37. Rodríguez, *Simón Rodríguez* 1:468.
38. Rodríguez, *Simón Rodríguez* 1:468.
39. For more on Marx's concerns regarding deforestation, see Ricardo Dobrovolski, "Marx's Ecology and the Understanding of Land Cover Change," 31–39.
40. Rodríguez, *Simón Rodríguez* 1:468.

41. Rodríguez, *Simón Rodríguez* 1:422–23.
42. Rodríguez, *Simón Rodríguez* 1:469.
43. Lilienthal, *TVA*, 2.
44. Rivera Garza, *Autobiografía del algodón*, 199.
45. Rodríguez, *Simón Rodríguez* 1:438.
46. Rodríguez, *Simón Rodríguez* 1:439.
47. Rodríguez, *Simón Rodríguez* 1:439.
48. Rodríguez, *Simón Rodríguez* 1:448.
49. Rodríguez, *Simón Rodríguez* 1:447–48.
50. Rodríguez, *Simón Rodríguez* 1:449.
51. Rivera Cusicanqui, *Ch'ixinakax utxiva*, 19.
52. Schwarz, *Misplaced Ideas*.
53. Rodríguez, *Simón Rodríguez* 2:505.
54. Cadena, "Indigenous Cosmopolitics," 359.
55. Molle, *Water Alternatives*, 332.
56. Paerregaard and Ullberg, *Water International*; Ullberg, "Making the Megaproject," 505. Zapana Churate has used a similar phrase, "un ambiente hidrosocial" (hydrosocial atmosphere), in the context of Majes Siguas. Efraín and Churate, "Respuestas a la crisis hídrica," 146.
57. Rodríguez, *Simón Rodríguez* 1:466.
58. Cant, "'Land for Those Who Work It,'" 1–37.
59. Rodríguez, *Simón Rodríguez* 1:466.
60. Rodríguez, *Simón Rodríguez* 1:423.
61. Olavarría Araya, "Prólogo Galeato," 62.
62. Stensrud, "Safe," 83.

BIBLIOGRAPHY

"Acta de asociación del Estado y particulares." 10 Mar. 1826. Archivo Regional de Arequipa. Protocolo 723. Notario Pedro de Luque. pp. 318–24.

Andersen, Astrid Oberborbeck. "Of Volcanoes, Saints, Trash, and Frogs: Eschatalogical Talks and Plural Ecologies in Arequipa, Peru." In *A Non-Secular Anthropocene: Spirits, Specters and Other Nonhumans in a Time of Environmental Change*, edited by N. Burbant, 31–45. More-than-Human. AURA Working Papers, vol. 3. Department of Culture and Society, Aarhus University, 2018.

Barriga, P. Víctor M. *Memorias para la historia de Arequipa*, vol. 1. La Colmena, 1941.

Basadre, Jorge. *Historia de la República del Perú*, vol. 1, *1822–1866*. Editorial Cultura Antártica, 1949.

Baudin, Luis. *El imperio socialista de los Incas*, 5th ed. Translated by José Antonio Arze. Zig-Zag, 1962.

Belaunde, Víctor Andrés. *Memorias*, vol. 1. Lima, 1960.

Briggs, Ronald. *Tropes of Enlightenment in the Age of Bolívar: Simón Rodríguez and the American Essay at Revolution*. Vanderbilt University Press, 2010.

Cadena, Marisol de la. "Indigenous Cosmopolitics in the Andes: Conceptual Reflections

Beyond 'Politics.'" *Cultural Anthropology* 25, no. 2 (2010): 334–70. DOI: 10.111/j.1548-1360.2010.01061.x.

Cant, Anna. "'Land for Those Who Work It': A Visual Analysis of Agrarian Reform in Velasco's Peru." *Journal of Latin American Studies* 44, no. 1 (Feb. 2012): 1–37. jstor.org/stable/41349718.

Chambers, Sarah C. *From Subjects to Citizens: Honor, Gender, and Politics in Arequipa, Peru, 1780–1854.* Pennsylvania State University Press, 1999.

Condori, Víctor. "Economía y empresa en Arequipa a inicios de la República, 1825–1850." *Economía* [Lima] 37, no. 74 (Jul.–Dec. 2014): 163–212.

Dobrovolski, Ricardo. "Marx's Ecology and the Understanding of Land Cover Change." *Monthly Review* 64, no. 1 (May 2012): 31–39.

Foster, John Bellamy. *Marx's Ecology: Materialism and Nature.* Monthly Review Press, 2000.

French, Jennifer. "Voices in the Wilderness: Environment, Colonialism, and Coloniality in Latin American Literature." *Review: Literature and Arts of the Americas* 45, no. 2 (2012): 157–66. DOI: 10.1080/08905762.2012.719766.

Gelles, Paul H. *Water and Power in Highland Peru: The Cultural Politics of Irrigation and Development.* Rutgers University Press, 2000.

Hurd, H. C. "Estudiar para aumentar las aguas del Río Chili (Arequipa)." Boletín del Cuerpo de Ingenieros de Minas del Perú, no. 34 (1905): 7–43.

Larkin, Brian. "The Politics and Poetics of Infrastructure." *Annual Review of Anthropology,* 42 (2013): 327–43. https://www.jstor.org/stable/43049305.

Leguía y Martínez, Germán. *Historia de Arequipa.* 2 vols. Lima: Impresa Moderna, 1912–1914.

Lilienthal, David E. *TVA: Democracy on the March.* Harper and Brothers, 1944.

López Palma, Jorge. *Simón Rodríguez: Utopía y socialismo.* Cátedra "Pío Tamayo," 1989.

Love, Thomas F. *The Independent Republic of Arequipa.* University of Texas Press, 2017.

Lozano y Lozano, Fabio. *El maestro del Libertador.* P. Ollendorff, 1913.

Lustgarten, Abraham. "The Colorado Is Drying Up Fast." *New York Times,* Aug. 29, 2021.

Manheim, Bruce, and Guillermo Salas Carreño. "Wak'as: Entifications of the Andean Sacred." In *The Archaeology of Wak'as: Explorations in the Sacred in the Pre-Columbian Andes,* edited by Tamara L. Bray, 47–72. University Press of Colorado, 2015.

Marx, Karl, and Frederick Engels. "Marx to Engels in Manchester [London,] 25 March 1868." Vol. 42 of *Karl Marx and Frederick Engels: Collected Works,* 557–59. Translated by Richard Dixon and others. International Publishers, 1975.

Molle, François, Peter P. Mollinga, and Phillipus Webster. "Hydraulic Bureaucracies and the Hydraulic Mission: Flows of Water, Flows of Power." *Water Alternatives* 2, no. 3 (2009): 328–49. wateralternatives.org.

Olavarría Araya, Braulino. "Prólogo Galeato a nombre de Simón Rodríguez." *Teré: Revista de filosofía y socio política de la educación* 3, no. 6 (2007): 59–73.

Quiroz, Alfonso W. *Corrupt Circles: A History of Unbound Graft in Peru.* Woodrow Wilson Center Press / Johns Hopkins University Press, 2008.

Paerregaard, Karsten, and Susann Baez Ullberg. "Smooth Flows? Hydrosocial Communities, Water Governance and Infrastructural Discord in Peru's

Southern Highlands." *Water International* 41, no. 3 (May 2020): n. pag. DOI: 10.1080/02508060.2020.1755538.

Polar, Jorge. *Arequipa: Descripción y estudio social.* Tip. Mercantil, 1891.

Pulgar Machado, Camila. *La materia y el individuo: Estudio literario de "Sociedades Americanas" de Simón Rodríguez.* Fundación Editorial Perro y Rana, 2006.

Regal, Alberto. *Los trabajos hidráulicos del Inca en el antiguo Perú,* 2nd ed. Instituto Nacional de Cultura, 2005.

El Republicano [Arequipa], April 8, 1826.

Rivera Cusicanqui, Silvia. *Ch'ixinakax utxiva: Una reflexión sobre prácticas y discursos colonizadores.* Tinta Limón, 2010.

Rivera Garza, Cristina. *Autobiografía del algodón.* Penguin Random, 2020.

Rivero y Ustáriz, Mariano Eduardo de, and Clemente Althaus. "Reconocimiento de la obra de Vincocaya." In *Colección de memorias científicas, agrícolas é industriales,* by Mariano Eduardo de Rivero y Ustáriz, 268–74. H. Goemaere, 1857.

Rodríguez, Simón. *Simón Rodríguez: Obras completas.* 2 vols. Ediciones del Congreso de la República, 1988.

Romaña, Eduardo L. de. "Informe del ingeniero Don Eduardo L. de Romaña." [1889]. *Anales de las obras públicas del Perú, Año 1899,* 558–56. Torres Aguirre, 1900.

Rozitchner, León. *Filosofía y emancipación: Simón Rodríguez: El triunfo de un fracaso ejemplar.* Biblioteca Nacional, 2012.

Rumazo González, Alfonso. *Simón Rodríguez, maestro de América: Biografía.* Universidad Simón Rodríguez, 1976.

Schwarz, Roberto. *Misplaced Ideas: Essays on Brazilian Culture,* edited by John Gledson. Verso, 1992.

Sohn-Rethel, Alfred. *Intellectual and Manual Labour: A Critique of Epistemology.* Translated by Martin Sohn-Rethel. Macmillan, 1978.

Stensrud, Astrid B. "Dreams of Growth and Fear of Water Crisis: The Ambivalence of 'Progress' in the Majes-Siguas Irrigation Project, Peru." *History and Anthropology,* 27, no. 5 (2016): 569–84. DOI: 10.1080/027572062016.1222526.

Stensrud, Astrid B. "Safe Milk and Risky Quinoa: The Lottery and Precarity of Farming in Peru." *Focaal-Journal of Global and Historical Anthropology* 83 (2019): 72–84. DOI: 10.3167/fcl.2019.831108.

Tamayo, A., et al. "Aguas del Chili." [1878] *Anales de las obras públicas del Perú, Año 1899,* 553–58. Torres Aguirre, 1900.

Ullberg, Susan Baez. "Making the Megaproject: Water Infrastructure and Hydrocracy at the Public-Private Interface in Peru." *Water Alternatives,* 12, no. 2 (2019): 503–20. wateralternatives.org.

Uriarte, Javier. *The Desertmakers: Travel, War and the State in Latin America.* Routledge, 2020.

Wibel, John Frederick. "The Evolution of a Regional Community Within Spanish Empire and Peruvian Nation: Arequipa, 1780–1845." PhD diss., Stanford University, 1975.

Zapana Churate, Luis Efraín. "Respuestas a la crisis hídrica en zonas agrícolas y urbanas: Caso de estudio 'Proyecto de Irrigación Majes Siguas I' Arequipa, Perú." *Agua y Territorio,* 12 (Jul.–Dec. 2018): 145–56. DOI: 10.17561/at.12.3532.

CHAPTER 2

Forests of Sound

Listening, Affect, and Matter in
Humboldt and Hudson

JENS ANDERMANN

La Selva (The jungle, 1998), Spanish sound artist Francisco López's seventy-minute compilation of "sound environments from a neotropical rainforest" made from field recordings in the homonymous Costa Rican lowland forest reserve during two successive rainy seasons, is on one of its many levels an aural pastiche referring back to one of the founding texts of acoustic ecology: Alexander von Humboldt's short essay "Das nächtliche Thierleben im Urwalde" (The nocturnal life of animals in the jungle).[1] Added to what was certainly the German naturalist's most popular work in his lifetime, *Ansichten der Natur* (Views of nature, 1808), only in its third edition published in 1849, Humboldt's short essay compels us to join its narrator in an act of immersion in the forest soundscape that gradually penetrates into, and replaces, the "clamor" of anonymous animal noises through a more structured listening. Following a lengthy and—just as its fluvial setting—somewhat meandering discussion on the relations between the intensity of the "feeling for nature" (*Lebendigkeit des Naturgefühls*) in the native inhabitants of forest and mountain habitats and its potential impact on linguistic expression, as well as the loss of a beloved dog, possibly to jaguar attack,

39

and the complications of the fluvial topography delaying a rescue mission in search of the canine friend, Humboldt's chapter eventually zooms in on the sleepless naturalists bivouacking in the upper Orinoco delta, unable to rest due to the unrelenting "cries of wild beasts thunder[ing] through the woods."[2] Always eager to turn adversity into opportunity, however, Humboldt quickly learns, aided as usual by his Indigenous guides, to break down this undifferentiated noise into individual sound patterns associated with particular animal species and to recreate their ebbs and flows in his own narrative tapestry. Or better: Humboldt invites us to read his text as a verbal score, an evocative notation of the rhythms, pitches, and timbres of animal voices that, just like individual instruments in an orchestral work, rise above the forest's incessant *basso continuo*. "Among the many voices that simultaneously gave cry," writes Humboldt,

> the Indians could identify only those that could be heard singly after a short pause. There were the monotonous, plaintive howls of the alouattae (howler monkeys), the whining, finely piping tone of the little sapajous, the quavering grumble of the striped night monkeys (*Nyctipithecus trivirgatus*, which I first described), the sporadic cries of the great tiger, the cougar or maneless American lion, the peccary, the sloth, and a host of parrots, parraquas (*Ortelida*), and other pheasantlike birds. . . . Occasionally, the tiger's cry would come down from the top of a tree. In these instances it would always be accompanied by the piping tones of the monkeys, who sought to escape this unusual pursuit.[3]

Humboldt's account of "the education of his senses"—in Oliver Lubrich's apt expression—also sketches out, and puts into experimental practice, an alternative project to *Views of Nature*'s overall, visual-ekphrastic framework of transcribing life into text.[4] Instead of mapping out a visual prospect (as, most emblematically, in the famous cross-section of the Andes accompanying Humboldt's *Essai sur la géographie des plantes*, 1805, where the writing itself is arranged to mimic the variations in vegetation according to altitude), "The Nocturnal Life of Animals in the Jungle" zeroes in on life's sonic manifestations. Here, interspecies and material "transfections," life-and-death relationships invisible to the eye, play themselves out on the level of sound.[5] The text gradually develops a technique of "close-listening," which, once established, will also be deployed by Humboldt on the only seemingly silent diurnal forest:

if one were to listen now, however, for the quietest tones that come to us in this apparent stillness of Nature, then one perceives close to the ground and in the lower layers of the atmosphere a muffled sound, a whirring and buzzing of insects. Everything announces a world of active, organic powers. In every shrub, in the cracked bark of trees, in the loose earth where live the hymenoptera, Life audibly stirs. It is one of the many voices of Nature, discernible to the solemn, receptive mind of humanity.[6]

It is as if the absence, in the nocturnal forest, of visual purchase on reality-as-landscape had suddenly alerted Humboldt to something that had always been co-present in, but also in excess of, landscape's apparatus of capture, hence requiring a different kind of mediation in order to be both sensually and rationally apprehended. As Lubrich points out, in Humboldt's account of nightly listening, "*Ansichten der Natur* have become *Stimmen der Natur*—voices of nature—as if Humboldt had spontaneously changed the character of his project. He has thus learned to convey the character of a place not solely as *vision* bus also as a *symphony*: *land*scape as *sound*scape."[7]

Aided by powerful digital recording technology, López's *La Selva* seems to finally realize what Humboldt could only invoke in the itself silent medium of writing—even though, as I shall argue further down, the naturalist's trans-scriptural routine as sketched out in "The Nocturnal Life of Animals" in fact comes quite close to being a field recording device in its own right. Compiled from individual fragments and edited into a temporal sequence imitating (in much-condensed fashion) "a protoypical day cycle of the rainy season beginning and ending at night," López's album-length sound piece brings back to mind Humboldt's analogy of the orchestral forest.[8] Beginning with a percussive chatter that may or may not be the composite sound of cicadas, frogs, and bird cries (but which also sounds a lot like electronic noise in industrial techno music), López's sound piece takes us through the *accelerandi* and *ralentandi* of multiple animal voices as well as their reverberations and those of nonorganic forces such as rain and thunder through the tree canopy and the underbrush. Long periods of relative stillness, alerting us to the occasional bird or insect sound piercing the low-level rhythmic tapestry of the cicadas, suddenly give way to the dramatic *crescendo* of thunder welling up from a distant murmur to full-blown banging on the timpani, and of storm agitating the treetops before torrential rain hits and literally drowns out all other voices.

Yet, all the while he pays homage to the German naturalist's foundational role in conceiving place as an aural rather than just a visual ensemble, López ultimately advocates for a different kind of listening than that imagined by Humboldt. For, even though López has at his disposal the very kinds of recording devices Humboldt can only anticipate in and through writing, he deploys these not in order to craft an aural tableau emulating (as well as complementing) the visual prospect of painterly landscape, with individualized animal voices stepping forth and occupying center stage before they fade back into the sonic background of life's incessant "whirring and buzzing." Rather, mobilizing French modernist composer Pierre Schaeffer's concept of "acousmatics"—"a sound that is heard without its cause or source being seen"[9]—López's sound work is after "a perceptual shifting from recognition and differentiation of sound sources to the appreciation of the resulting sound matter."[10] Acousmatic experience, rather than to soundscape's ekphrastic "sonic image," "can contribute significantly to . . . 'blindness' or profound listening."[11] To the aural extractivism of the soundscape –which, just as other modes of naturalist fieldwork, singles out and embalms individual "samples" or specimens *in situ*, to subsequently export these, as "immutable mobiles," to the "center of calculation" (the lab or recording studio) where they can be analyzed, edited and re-constellated—ecological acousmatics opposes an immersive listening in the sonic milieu, drawing on the listener's body as subject to affective interpellation and responses.[12]

In this chapter, I want to contrast Humboldt's soundscape writing with an early instance of the sonic milieu as a harbinger of affective immersion in place, unleashing a dissident geo- and anatomopolitics of sound. I will do this by pivoting to the opposite end of the nineteenth century and of the South American subcontinent, analyzing the work of Anglo-Argentine novelist, nature writer, ornithologist, and memorialist William Henry Hudson (1841–1922). Hudson's liminal presence in the English as well as Argentine literary canon, including his biography that inverts the archetypal itinerary of the New World naturalist in his own life-journey from late-colonial margin to imperial metropolis, also turns aural immersiveness into something of a "postcolonial affect" –one that works inside natural history disrupting, and deconstructing, the very apparatus of capture to which it nominally contributes. Building on Mary Louise Pratt's foundational reading of the Humboldtian "totalizing project [that] lives *in* the text, orchestrated by the infinitely expansive mind and soul of the speaker," my comparison of Humboldt's and Hudson's engagements with the aural as well as with the nonvisual environment more widely also attempts to complicate the notion of "anti-conquest" rhetoric she attributes both to Humboldtian natural history and to Anglo-American resource prospecting.[13] Although "the erasure of the human" is effectively a shared feature which compromises the politics of Humboldt's and (albeit to a lesser degree) Hudson's nature-writing, the very different positioning of the observer's body vis-à-vis the sonic rather than the visual environment also allows for a different, post-colonial or even post-extractivist, reading of some areas of their work.[14] Moreover, whereas critics such as Timothy Morton have taken Western "environmental awareness" to task over the disavowal of its own complicity with –or indeed reinforcement of– the exteriority and objectness of "nature" it purportedly decries, in Humboldt's and Hudson's soundings of forest and plain, nature-writing itself begins to corrode this same founding assumption.[15] Returning to these "nineteenth-century works that were fundamental to the rise of the environmentalist movements of the twentieth century," as Ida Marie Olsen has suggested, we can productively complicate the genealogy of Western environmentalism, recovering those instances where it turns in on itself and suggests ways that "Morton's 'rigorous distinction' between environmentalism and ecological awareness can be reconsidered and negotiated."[16]

The Humboldt Effect: Sound, Memory and the Impression of Nature

"The Nocturnal Life of Animals in the Jungle" was not the first time Humboldt had shifted his attention from the visual to the aural environment. In fact, already in the naturalist's much earlier account of his and Aimé Bonpland's South American journey, *Voyages aux régions équinoxiales du nouveau continent* (Travels to the equinoctial regions of the new continent, 1799–1804), he had begun to speculate on the possible interfaces between sound phenomena, bodies and the physical and atmospheric environment. Just as close attention to visual correspondences in the physical setting and their physiognomic description afforded the sciences a possibility for revealing the invisible forces structuring the ensemble, Humboldt suggested, so the *event* of sound—its passage through time and space as well as its perception by both the ear and the body more generally—might hold a key to understanding the mutually co-affective relationship between organic and inorganic, material and atmospheric elements of the landscape. Once again, it is when the travelers have set up camp and are preparing for the night, that the naturalist's attention switches from the visual to the aural register:

> We stopped near the raudal of Cunuri. The noise of the little cataract augmented sensibly during the night. Our Indians asserted, that it was a certain presage of rain. I recollected, that the mountaineers of the Alps have great confidence in the same prognostic. In fact, it rained long before sunrise, and the araguate monkeys had warned us by their lengthened howlings of the approach of the shower, long before the noise of the cataract.[17]

In a footnote to the same passage, Humboldt dismisses the French meteorologist Jean André Deluc's suggestion (from the latter's 1772 *Recherches sur les modifications de l'athmosphère*) that this change in sonic intensity might actually be the result of a change in barometric pressure leading to an increased presence of air bubbles bursting as they hit the water surface. Instead, he suggested that "the cause of the phenomenon is a modification of the atmosphere, which has an influence at once on the *sonorous* and on the *luminous undulations*. The prognostic drawn from the increase and the intensity of sound is intimately connected with the prognostic drawn from a lesser extinction of light."[18] Sound and light waves, Humboldt suggests, are mutually interconnected, *as are visual and aural impressions in the body that perceives them*: it is the embodied experience in the field, in other words, that provides Humboldt with a cue for his dissenting opinion from Deluc's about the greater and spatially more extensive perception of sound by night.

Subsequently, on his return to Paris, Humboldt would revisit the subject in 1820 in a lecture given at the French Academy of Sciences, entitled "Sur l'acroissement nocturne de l'intensité du son" (On the Increase of Sound Intensity at Night). Here, he refers once again to his experience, while sojourning at the Franciscan missions of Atures and Maypures in the Orinoco, of listening to the sound of cataracts "more than a league" away, although "when this noise is heard in the plain that surrounds the mission . . . you seem to be near a coast skirted by reefs and breakers. The noise is three times as loud by night as by day, and gives an inexpressible charm to these solitary scenes."[19] We should notice how, once again, it is aesthetic enjoyment as well as the difficulty of adequately rendering it in the medium of language—the solitary scenes' "inexpressible charm"—which leads on to scientific endeavor in order to get to the bottom of this intense attraction. Yet scientific insight is thus also constitutively reliant on the subjective, embodied sensation of affection by environmental stimuli, as Humboldt makes clear in moving from narrative description to speculative interpretation of the geochemistry of sound:

> The velocity of the propagation of sound, far from augmenting, decreases with the lowering of the temperature. The intensity diminishes also by dilation of the air, and is weaker in the higher than in the lower regions of the atmosphere, where the number of particles of air in motion is greater in the same radius. The intensity is the same in dry air, and in air mingled in vapours; but it is feebler in carbonic acid gas, than in mixtures of azot and oxygen. From these facts, which are all we know with any certainty, it is difficult to explain a phenomenon observed near any cascade in Europe, and which, long before our arrival in the village of Atures, had struck the missionary and the Indians.[20]

This is, of course, a description of the so-called Humboldt Effect, as the variation in the intensity of sound circulation during different hours of the day and across diverse geoclimatic zones has come to be known following the physicist Hans Ertel's suggestion in another lecture to the Academy of Sciences, this time in Berlin, more than a century after Humboldt's.[21] As the sound artist Daniel Velasco (director, among others, of the soundscape theater production *Humboldt & Bonpland*, 1999) argues, Humboldt's engagement with sound involves a complex back-and-forth between sound as a physical event, its perception and cognition by the senses and the body, and its memorization and transcription in the text. Environmental acoustics, in Humboldt's work, thus come about as a mobile, dynamic constellation between "four reference categories": "(1) hearing sounds, (2) remembering sounds, (3) sounds themselves, and (4) silence."[22] Thanks to his ability for combining "sensuous exactitude" with "an unusually sensitive and perceptive ability" in the registering, cognizing and describing of sounds, Velasco contends, Humboldt's "descriptions of sonic landscapes today impress us in their vivid reality, as his language took the place of the tape recorder in conserving sounds."[23]

Rather than a case of López's "blind" or profound listening, then, Humboldt's relationship with sound is always already entangled with the spatial and temporal activity of its classification and memorization, in which language and writing play a key part. Although physical immersion and affective opening of the listener's body toward the field is very much a condition of the Humboldtian soundscape, it is really as a mobile container for the collecting and on-site processing of raw data. In fact, written language has a twofold importance here, as it both re-arranges the soundscape in/as a textual tapestry by way of naming *and* returns to this transcription the original intensity of lived, bodily perception by way of narrative. Of course, we should keep in mind that this capacity for writerly recording of the soundscape relies, in fact, on multiple instances of translation both in the field itself and in the metropolitan center of calculation where this text/tape is being produced—a kind of "transculturation in reverse" where the liveliness and proximity of the vernacular name to its (sound) source is not so much overwritten as brought into the service of a new kind of metropolitan science.

Indeed, in "The Nocturnal Life of Animals" it is effectively not the naturalist but his Indigenous guides who first single out, and put a name to, the animal sound that is being recognized. This first name, produced through an act of collective memory-making, is the native or "vulgar" one that is still relatively close to the sound source itself, which (as in "parraqua") it may even mimic through the use of onomatopoeia. Humboldt's original German version of the essay does something quite similar in the way it proceeds to transcribe this native term into a German noun—"*aluates*" becomes "*die Aluaten*"—after which, in parenthesis, are usually added either the German vernacular name or its Latin taxonomic equivalent. Moreover, each of these double acts of naming is also accompanied by a short description of the sounds emitted by each of the animals thus identified, generally by likening them to musical instruments or to vocal timbres. What Humboldt is inventing here, as suggested by Velasco, is a recording technology *avant-la-lettre* (or perhaps rather: *avant-le-machine*)—a methodology of ekphrastic note-taking that approximates in the medium of writing the capture and classification of sounds that would only become possible some thirty years later through Thomas Edison's phonographic cylinders.

For "noise" to turn into "voices of nature" revealing themselves to "the pious and susceptible spirits of man" then, it must also undergo a multi-step process of translations, first in the field and then at the desk. All of these steps are necessary as "nature" must reveal itself to the human "spirit," in equal measure, through feeling and through reason: hence, the language of classification must never overwrite but merely bestow order on, the original, physical sensation of listening. Humboldt's point in "The Nocturnal Life of Animals" as well as in *Views of Nature* as a whole is, in fact, twofold. On the one hand, he is reflecting about the differences between nature's visual apprehension as landscape or prospect and as a sonic and rhythmic texture that makes manifest to the sensorium—indeed, turns the human body into *an instrument* resonant with—the interplay of living forces. Yet on the other hand, framing this discussion of the differences between visual and aural purchase on one's surroundings, there is also a more general argument concerning the relation between language and the senses—that is, the degree of proximity to and of detachment from, the *Naturgefühl* or "feeling of nature" as opposed to the latter's abstraction and capture by way of taxonomic classification. The wider question Humboldt is after is about how erudite, scientific language can hold on to the felt "liveliness" of natural elements that remains present, he asserts, in the native languages of the inhabitants of steppes, deserts and jungles. Indigenous languages, Humboldt claims, literally bear the imprint of close, daily contact with nonhuman organisms and materialities and, thus, remain concerned with use rather than exchange value, with the hunter-gatherer's need for interspecies channels of communication rather than the naturalist's abstract, orderly naming of living organisms. "The speech of humans is enlivened by everything indicative of natural truth," writes Humboldt, "be it in the representation of sensory impressions received from the outer world or of profoundly stirred thought and inner feelings."[24]

But then, language itself is in fact a crossroads of "animations," a kind of membrane that is permeable from both sides. Language is an exchange medium between environmental "impressions," on the one hand, and the ideas and feelings emerging from inside the mind, on the other. The trick of naturalist description—not unlike that of shamanic invocation—is to facilitate this in- and outflow through a technique of controlled suspension of thought. "That which is written down on the spot," Humboldt claims, "or soon after the impression of the phenomena has been received, may at least proclaim to possess more freshness [*Lebensfrische*] than what is produced by the recollection of long past events."[25] The art of writing in the field is to preserve the plasticity of the words received from an environment's native inhabitants. This requires an exercise of self-limitation on the part of the observer, in order to maximize the mind's permeability as a kind of embodied recording device. Making language amenable to the "impressions" of the location, Humboldt concludes "is most easily achieved by simple narration of what has been observed and experienced directly, through the limiting individualization of the situation on which the narrative hangs."[26] "Individualization of the situation": a kind of proto-cinematic montage, performed in and through writing, from wide-angled panorama to close and medium-range POV shots in which the lived "situation" and its affective "impression" are recovered by switching from description to narrative and from the armchair scientist's detached distance to the field naturalist's physical and affective immersion.

By subsequently moving from these general considerations on writing in the field to the "sample case" of the nocturnal forest transcribed into a textured as well as textualized soundscape, Humboldt also appears to single out sound, rather than vision, as a shortcut from life to language. Because language itself is sonic, the insistent stirring [*Regung*] of the living which, in the final scene of the naturalist pressing his ear to the bark of a tree, manifests the ever-present interplay of the great vital forces, also remains *materially present* in the linguistic sign, in much more vivid fashion than ocular impressions in the landscape view. For Humboldt, soundscape has to supplement landscape for the latter to be able to convey a "living image." The space and time of sound complement those of vision because they exceed, rather than coincide with, the visual *tableau*. Therefore, at night, when vision is suspended and the naturalist does not have to busy himself with the exercise of visual capture and composition, he can at last lend his ear to the sonic matter all around him. The "voices of nature" are the supplement emerging in the space and time of the suspended image: exactly the position that "The Nocturnal Life of Animals" occupies within *Views of Nature* as a whole.

Adventures Among Birds: Sensory Immersiveness in W. H. Hudson

As so many of the little vignettes associating the description of a particular animal or plant species with the time, space, and affective experience of its encounter, which are a kind of narrative base unit in W. H. Hudson's work, the entry on the white-banded mockingbird (*Mimus triurus*) in *Birds of La Plata* (1920)—an expanded and updated edition of his own *Argentine Ornithology*, published in two volumes in 1888–89—is at the same time a veiled autobiographical fragment and a critique, in the name of an animist aesthetics, of scientific apprehension of nonhuman nature, advocating instead for a "language of passion."[27] Hudson notes that the Spanish naturalist Azara was the first to identify this bird species in Paraguay, but his assertion that it possessed no melodious notes "proves at once that he never heard it sing," just as later authors including D'Orbigny and Bridges "have told us nothing of its song and of its miraculous mocking powers."[28] What the Spanish, French and British naturalists all missed was, in fact, the very essence of the bird also known in the region as "calandria blanca" since, as "the natives all assured me, it possessed a very wonderful song, surpassing the songs of all other birds."[29] Just as in Humboldt's field-recording episode on animal sounds in the jungle, in Hudson's Patagonian vignette scientific argument switches to first-person narrative in order to make a point that is crucially reliant on this embodied, sensory experience in the field. "While walking through a chañar-wood one bright morning," Hudson tells us,

> my attention was suddenly arrested by notes issuing from a thicket close by, to which I listened in delighted astonishment, so vastly superior in melody, strength, and variety did they seem to all other bird-music. That it was the song of a *Mimus* did not occur to me; for while the music came in a continuous stream until I marvelled that the throat of any bird could sustain so powerful and varied a song for so long a time, it was never degraded by the harsh cries, fantastical flights, and squealing buffooneries so frequently introduced by the Calandria, but every note was in harmony and uttered with a rapidity and joyous abandon no other bird is capable of, except, perhaps, the Skylark; while the purity of the sounds gave to the whole performance something of the ethereal rapturous character of the Lark's song when it comes to the listener from a great height in the air.[30]

Enchanted and mystified by the strange and unexpected concert, the narrator remains quiet and motionless for fear of scaring away the unknown vocalist, when suddenly a different voice—"the shrill, confused, impetuous song of the small Yellow-and-Grey Patagonian Flycatcher"—interrupts the song.[31] Frustrated that he might never learn the identity of his entertainer, he moves forward when,

> from the same spot, came the mellow mating-song of the Diuca Finch, and this was quickly succeeded by the silvery bell-like trilling song of the Churrinche, or little Scarlet Tyrant-bird. Then followed many other familiar notes and songs —the flute-like evening call of the Crested Tinamu, the gay hurried twittering of the Black-headed Siskin, and the leisurely-uttered delicious strains of the Yellow Cardinal, all repeated with miraculous fidelity. How much was my wonder and admiration increased by the

discovery that my one sweet singer had produced all these diverse strains! The discovery was only made when he began to repeat songs of species that never visit Patagonia. I knew then that I was at last listening to the famed White Mocking-bird, just returned from his winter travels, and repeating in this southern region the notes he had acquired in sub-tropical forests a thousand miles away.[32]

Finally catching sight of the animal in time to admire it being "carried away with rapture" as it "resumed [its] own matchless song," the narrator can only conclude that "this bird is among the song-birds like the diamond among stones, which in its many-coloured splendour represents and exceeds the beauty of every other gem."[33] In fact, he adds, he has since been able to confirm his belief as, "shortly after hearing it, I visited England, and found of how much less account than this Patagonian bird, which no poet has ever praised, were the sweetest of the famed melodists of the Old World."[34]

Hudson's Patagonian birding story, as we see at once, is very similar in textual structure to Humboldt's immersion in the nightly jungle sound-scape. Both start with a general, impersonal discussion of particular questions in the natural sciences, only to then switch register and launch into a first-person narrative of embodied experience in the field. Abstract, detached description is submitted in both texts to the corrective of embodied, sensorial experience, including the observer's own affective responses, to the effect of "a minimization of the distance between observer and object of study."[35] Yet whereas, for Humboldt, this instance of embodied, subjective observation, is but the raw material for the speculative production of objective knowledge upon the traveler's return to the metropolitan center—akin to the lab experiment, as Oliver Lubrich and Ottmar Ette point out, but with the naturalist's own body as simultaneously the subject of trial and measuring device—Hudson's mode of immersiveness is of a different and, I would argue, also a less colonial-extractive kind.[36] "If there is anything one feels inclined to abhor in this placid land," he writes in *Idle Days in Patagonia*, "it is the doctrine that all our investigations into nature are for some benefit, present or future, to the human race."[37] At the same time, the very possibility of distilling a detached, objective kind of knowledge by reverting from the first to the third person and from embodied experience to generalized description, is called into question as, in the field, the naturalist is but one body among many, whose presence affects the ensemble just as he is affected by the environment: "no bird, conscious of being watched, will act unconstrainedly any more than a human being with clouded reputation will comport himself naturally with the eyes of a detective on him. While we are observing the bird, the bird watches us; of all its curious things when we are out of sight and mind, we see nothing."[38]

There can be no "observation of nature" that is not itself already coextensive with what is being observed, or as Hudson puts it in an oft-cited passage from *Hampshire Days*: "The blue sky, the brown soil beneath, the grass, the trees, the animals, the wind, and rain, and sun, and stars are never strange to me; for I am in and of and am one with them; and my flesh and the soil are one, and the heat in my blood and in the sunshine are one, and the winds and tempests and my passions are one."[39] This monistic vision of an enmeshed, "sympoietic" universe "always partnered all the way down, with no starting and subsequently interacting 'units,'" in Donna Haraway's expression, is also in evidence in Hudson's encounter with the white-banded mockingbird, which—and let us remember, within a book that is ostensibly an ornithological encyclopedia—enacts a double challenge to the notion of objective science and to its colonial underpinnings.[40] First and foremost, Hudson's own itinerary from colonial margin to imperial center challenges the very geopolitics of knowledge that organizes the field of natural history, as it is England that is being "visited" here by the overseas naturalist and turned into a site of "fieldwork": the site of "culture" is turned into "nature," and an inferior one at that, compared with the beauty and variety of the author's native land, Argentina. Yet at the same time, and far more radically, Hudson's nature-writing from the margins does not so much attempt to better the Old World poets' praise for less talented songbirds with his own praise for the calandria blanca, as he credits the animal itself with being able to do what human language cannot, since "bird music, and, indeed, bird sounds, generally, are seldom describable. We have no symbols to represent such sounds on paper, hence we are powerless to convey to another the impression they make on us as we are to describe the odour of flowers."[41] The embodied experience of affectedness by bird music (which is not the same as emotion) is in tension with language, which can only invoke the affect but not the sound that triggered it. The calandria, meanwhile, is able to reproduce "with miraculous fidelity" the songs of local as well as exotic birds it has listened to in its travels, thanks to its greater capacity for immersive incorporation, for "minimizing the distance" between the representation and its object—its capacity for becoming-Churrinche, or Crested Tinamu, or Yellow Cardinal.

Just as there is no "outside" of nature, then, its representation is also no prerogative of the human species: the mimetic faculty, and even aesthetic experience, are common to all living forms, Hudson argues, and not just for reasons of sexual selection or camouflage—as would claim Darwin or the entomologist Henry Walter Bates, whose pioneering research on insect mimicry Hudson admired and drew upon in his fiction—but for its own sake. Beauty is the effect of an ecumenic, joyful experience of community with the living, the "periodical fits of gladness" all living things are prone to, "affecting them powerfully and standing out in vivid contrast to their ordinary temper. And we know what this feeling is—this periodic intense elation, which even civilized man occasionally experiences," as Hudson writes in a chapter on "Music and Dancing in Nature" in *The Naturalist in La Plata*.[42]

Whereas, for Humboldt, the intensity of affective, embodied experience in the field is but a conduit toward a form of writing that, whilst retaining its freshness and vitality [*Lebensfrische*], also subjects the data thus gathered to analytical speculation after-the-fact, Hudson's writing seems to move in the opposite direction. In his nonfiction works, what looks at first to be another (popular) treatise on natural history,[43] eventually turns out to be a particular kind of memory exercise striving to recover those moments of "free[dom] from apprehension . . . as if I had changed my identity for that of another man or animal," as Hudson himself puts it in *Idle Days in Patagonia*.[44] It is the moments of becoming-animal, of re-immersion into the mesh of existents, which turn out to be the focal point of his nature writing and his fiction alike. These moments of *becoming*, where the "idling mind" gains textual form as if the narrator had been "pressing the 'pause button'"—in Heidi Liedke's observation—are central to Hudson's nature-writing, as it pivots away from description of the physical world and toward a "virtual landscape" of enmeshment and sympoieisis, where the boundaries between self and environment have collapsed.[45] Idling, or "freedom from apprehension," in Hudson, is very similar, in fact, to what López in his plea for an acousmatics of sonic matter calls "blind listening": a yielding, nonpurposeful relation between the senses and the object of perception that ultimately aims at exploding this very boundary toward a new, third space of becomings. The climactic instances in Hudson's work are the ones in which the apprehensive mind is successfully tricked into inaction and observation turns into immersion, such as when, while resting in the midst of the desert and "listening to the silence . . ., animal forms did not cross my vision or bird voices assail my hearing . . . In that novel state I was in, thought had become impossible."[46]

"Listening to the silence"—a state of openness, of yielding to the body of sound even and especially in the latter's absence—provides access to a kind of wholeness that, for all its affinities, is the opposite of Humboldt's totalizing project of sensory-analytical capture. It is an experience that, as Jean-Philippe Barnabé describes it, "translates the deliberate and joyous emptying-out of a subject that has suspended all physical effort, silenced any kind of interior discourse, and which strives to let go of its singularity and its temporal condition to find shelter in an all-envolving cosmic totality, integrated into the pure flow of a perpetual present."[47] What Hudson variously describes as his animism—"not a doctrine of souls that survive the bodies and objects they inhabit, but the mind's projection of itself into nature, its attribution of its own sentient life and intelligence to all things"—is a nonextractive, nonpurposeful relation with the living and the inorganic that is also closely related, paradoxically, with the author's own liminal identity: his own "classificatory instability" on the edge of disciplines, styles, and nationalities.[48] Immersiveness comes out of a radical non-belonging, which incessantly turns Hudson's writing into a third space, a space of constant becoming that is neither literature nor science, neither English nor Argentine, and where thanks to this very instability of the subject of enunciation a more permeable, multisensory and even—to use Eduardo Viveiros de Castro's expression—"perspectivist" or "multinaturalist" attitude toward the mesh of existents becomes possible.[49] "To know the creature, undivested of life or liberty, or of anything belonging to it—Hudson suggests in *The Book of a Naturalist*—it must be seen with an atmosphere, in the midst of nature in which it harmoniously moves and has its being, and the image it casts on the observer's retina and mind must be identical with its image in the eye and mind of other wild creatures that share the earth with it."[50] Only by seeing like an animal does one see the animal, Hudson suggests, and this "atmosphere" one has to enter (which is but another way of referring to the immersive, trance-like state of becoming-other) is more easily available to hearing than to vision, making sound beings—such as birds—a central concern of Hudson's work. The temptation confessed at the outset of *Far Away and Long Ago* "to make this sketch . . . a book about birds and little else" is in fact a constant in Hudson's entire oeuvre.[51]

Finale: Sound as Becoming-With

Hearing, as cultural geographer Yi-Fu Tuan argues, is closer to touch than vision, not least because, as listeners, we are on the receiving end rather than in charge of the action.[52] Unlike visual landscape, which renders our surroundings into an object at the behest of our gaze, the aural environment undermines subjective self-entitlement and instead favors "consubstantiation with nature."[53] On the level of writing, this change of ekphrastic register from visual landscape to soundscape has the effect of calling forth an experimental and highly self-reflexive kind of writing that, while it tries out and comments on different ways of transcribing and translating nonhuman sounds, is also exceptionally alert to the affective intensities that sonic patterns trigger in the listener's body, making it resonate with its surroundings.

For Humboldt, the importance of these "Stimmen der Natur" (voices of nature) is at least twofold. On the one hand, as we have seen, the soundscape offers him not just an alternative possibility to visual description for identifying animal species and their mutual relationships within a particular habitat. Moreover, and just as importantly, sound also opens up a field of translatability from Indigenous languages to scientific description, as a way of preserving (in the form of a Latourian "immutable mobile") the vitality of the object of description in the name that describes it. Vernacular languages display a greater proximity to their living referent on account of their—physical as well as phonetic—closeness to nonhuman sonorities, which is why (unlike in the clear-cut division between the beholder and his vision) the relationship between a sound source and its listener is one of successive instances of transculturation, rearranging and repurposing nonhuman voicings and their vivid "impressions" in Indigenous vernaculars for a new kind of "life science." Secondly, the varying intensities of sound in space and time also provide Humboldt with an opportunity for thinking together the living and the inorganic. Since sound is at once the product and object of perception of living bodies and is mediated in its power to affect the receiver by the geoclimatic setting in which it occurs, it is also a key example for the way in which "invisible forces of life" run through organic and geodesical bodies alike, structuring both in their mutual inter-affectedness. The description of the soundscape, as in "The Nocturnal Life of Animals," by way of the orchestral or choral analogy that makes the text itself into a kind of verbal score, complements the writerly evocation of visual landscape in its capacity to tease out the invisible—and inaudible—forces that govern the ensemble. Humboldt's "total landscape" writing, as Oliver Lubrich calls it, represents a crossroads of ekphrastic transcriptions, it "is . . . at once optical and acoustic, pictorial and poetic—it becomes multimedial."[54]

In a chapter on "Bird Music in South America" included in *Idle Days in Patagonia*, Hudson compares his experience of wandering through English woodland with that of roaming across the plains and forests of his native Argentina. Just as Humboldt, Hudson resorts to musical analogies to make the case for the superior ability of "the melodists of my country" to affect the listener, especially if one considers "the character and value of the music" rather than just attempting to classify individual voices:[55]

> The bird-language of an English wood or orchard, made up in the most part of melodic tones, may be compared to a band composed entirely of small wind instruments with a limited range of sound, and which produces no storms of noise, eccentric flights, and violent contrasts, nor anything to startle the listener—a sweet but somewhat tame performance. The South American forest has more the character of an orchestra, in which a countless number of varied instruments take part in a performance in which there are many noisy discords, while the tender spiritual tones heard at intervals seem, by contrast, infinitely sweet and precious.[56]

Let us note here that this more "orchestral" character of forests and prairies as compared to the chamber ensemble of European countrysides is not just the effect of its richer panoply of animal voices; rather, for Hudson, "all natural sounds produce agreeable sensations . . . : the patter of rain on the forest leaves, the murmur of the wind, the lowing of the kine, the dash of waves on the beach."[57]

Although, then, Hudson's orchestral forest bears remarkable similarities to Humboldt's, there are also important differences. First and foremost, the affective impact of sound through bodily immersion of the listener here is not just, as it was for Humboldt, the necessary condition for its inspired transcription on the spot, to be subsequently analyzed and filtered out in the spatio-temporal distance of the study whilst retaining the vitality of the initial moment. Johannes Fabian, in the context of anthropological fieldwork, has called this double session of writing the "denial of coevalness, which allows a horizontal and dialogical encounter to be turned into an objectified otherness."[58] For Hudson, on the contrary, affect is itself the very matter of sound as it flows from one body to the other: it is not the context but the very text of sound, which has no other existence than this affective intensity between and among bodies.

Secondly, and if this is the case, sounds also never appear individually, one by one, but always as part of a bigger ensemble—a sonic milieu that comprises not just "animal voices" but also their interplay with the geophonies, or sound phenomena provoked by rainfall or the percussive effects of wind rushing through trees and high grasses. Just as in López's *La Selva*, in Hudson's evocation of the South American orchestral forest "the presence of the noisy milieu/medium is not minimized. Rather, signal and noise, foreground and background, event and context are presented together, alluding to the notion that what is heard stems from the combination of sound source and its environment."[59] Therefore, too, the individuation of sounds (which, just as the analogy with musical instruments and vocal registers, is a necessary scaffolding for their recreation in textual form), does not result in a soundscape the way it does in Humboldt. Sound as invoked in Hudson's writing does not produce the aural equivalent of Humboldt's typographic Cordillera from his essay on plant geography. Rather than being situated *in space*, for Hudson the transfection among sonic materialities is—or rather, *becomes*—the very space of immersion his writing takes us into. As Deleuze and Guattari would say, we are thus ushered into "a sonorous landscape [which] takes as its counterpoint all the relationships with a virtual landscape," a space of pure becoming.[60] This "virtual landscape," I have argued, is at the very limit of description toward which Hudson's writing constantly strives, and from which it receives at the same time its driving impulse: the space and time of a becoming-with—of "sympoiesis," of compostism—when "nature" ceases to be an object of description and instead the shared "fleshiness" of all things, living as well as inorganic, comes to the fore.

NOTES

1. I am using the English chapter title as rendered in Élise Otté's and Henry G. Bohn's first English translation of Humboldt's text, published in 1850. Mark W. Person's more recent version from 2014 has it down as "The Nocturnal Wildlife of the Primeval Forest," which doesn't quite catch the importance in Humboldt's soundscape of *animals* as sources of sound.
2. Humboldt, *Views of Nature*, 146.
3. Humboldt, *Views of Nature*, 146.
4. Lubrich, "Humboldtian Landscapes," 97.
5. Haraway, *Staying With the Trouble*, 1.

6. Humboldt, *Views of Nature*, 147.
7. Lubrich, "Humboldtian Landscapes," 97.
8. López, "Blind Listening," 163.
9. Chion, *The Voice in Cinema*, 18.
10. López, *La selva*, 1.
11. López, *La selva*, 2.
12. Latour, *Science in Action*, 236–43.
13. Pratt, *Imperial Eyes*, 125.
14. Pratt, *Imperial Eyes*, 125.
15. Morton, *Dark Ecology*, 129–36.
16. Olsen, "Outlines of Ecological Consciousness," 194–95.
17. Humboldt, *Personal Narrative*, 436–37.
18. Humboldt, *Personal Narrative*, 436.
19. Humboldt, *Kleinere Schriften*, 241.
20. Humboldt, *Kleinere Schriften*, 242.
21. Ertel, "Ein Problem der meteorologischen Akustik."
22. Velasco, "Island Landscape," 23.
23. Velasco, "Island Landscape," 23–24.
24. Humboldt, *Views of Nature*, 192.
25. Humboldt, *Views of Nature*, 192.
26. Humboldt, *Views of Nature*, 142.
27. Hudson, *Idle Days in Patagonia*, 79.
28. Hudson, *Birds of La Plata*, 10.
29. Hudson, *Birds of La Plata*, 11.
30. Hudson, *Birds of La Plata*, 11.
31. Hudson, *Birds of La Plata*, 12.
32. Hudson, *Birds of La Plata*, 13.
33. Hudson, *Birds of La Plata*, 10.
34. Hudson, *Birds of La Plata*, 13.
35. Olsen, "Outlines of Ecological Consciousness," 206.
36. Lubrich and Ette, "Versuch über Humboldt," 37.
37. Hudson, *Idle Days*, 92.
38. Hudson, *Birds of La Plata*, 339.
39. Hudson, *Hampshire Days*, 47.
40. Haraway, *Staying With the Trouble*, 33.
41. Hudson, *Idle Days*, 97.
42. Hudson, *The Naturalist in La Plata*, 280.
43. Heidi Liedke reminds us that *Idle Days in Patagonia* was written during a period when Hudson was experiencing intense financial hardship, which may have made the prospect of plugging into the contemporary craze for popular nature writing –while also drawing on his own memories from "exotic" environments, rather than engaging in time-consuming fieldwork– a promising one. See Liedke, *The Experience of Idling*, 146–47.
44. Hudson, *Idle Days*, 133.
45. Liedke, *The Experience of Idling*, 159; Deleuze and Guattari, *Mille Plateaux*, 392.
46. Hudson, *Idle Days*, 133.
47. Barnabé, "'Staring at Vacancy,'" 71.

48. Hudson, *Idle Days*, 79; Olsen, "Outlines of Ecological Consciousness," 196.
49. Castro, *Metafísicas canibais*, 22, 55.
50. Hudson, *The Book of a Naturalist*, 150.
51. Hudson, *Far Away and Long Ago*, 62.
52. Tuan, *Topophilia*.
53. Barnabé, "'Staring at Vacancy,'" 66.
54. Lubrich, "Humboldtian Landscape," 98.
55. Hudson, *Idle Days*, 101.
56. Hudson, *Idle Days*, 103–4.
57. Hudson, *Idle Days*, 103.
58. Fabian, *Time and the Other*, 35.
59. Thompson, *Beyond Unwanted Sound*, 88.
60. Deleuze and Guattari, *Mille Plateaux*, 391.

BIBLIOGRAPHY

Barnabé, Jean-Philippe. "'Staring at Vacancy': Notas sobre *Far Away and Long Ago*." In *Entre Borges y Conrad: Estética y territorio en William Henry Hudson*, edited by Leila Gómez and Sara Castro-Klarén, 65–80. Iberoamericana-Vervuert, 2012.

Castro, Eduardo Viveiros de. *Metafísicas canibais: Elementos para uma antropologia pós-estrutural*. Cosacnaify, 2015.

Chion, Michel. *The Voice in Cinema*. Translated by Claudia Gorbman. Columbia University Press, 1999.

Deleuze, Gilles, and Félix Guattari. *Mille Plateaux: Capitalisme et schizophrénie II*. Minuit, 1980.

Ertel, Hans. "Ein Problem der meteorologischen Akustik (Die tagesperiodische Variation der Schallintensität." Sitzungsberichte der Deutschen Akademie der Wissenschaften zu Berlin. Klasse für Mathematik, Physik und Technik 2. Berlin (Ost): Akademie-Verlag, 1955.

Fabian, Johannes. *Time and the Other: How Anthropology Makes Its Object*. Columbia University Press, 1983.

Haraway, Donna J. *Staying With the Trouble. Making Kin in the Chthulucene*. Duke University Press, 2013.

Hudson, William Henry. *Birds of La Plata*. J. M. Dent & Sons, 1920.

Hudson, William Henry. *The Book of a Naturalist*. Hodder & Stoughton, 1919.

Hudson, William Henry. *Far Away and Long Ago: A Childhood in Argentina* [1918]. Century Publishing, 1985.

Hudson, William Henry. 1903. *Hampshire Days*. London: J. M. Dent & Sons.

Hudson, William Henry. 1893. *Idle Days in Patagonia*. Stroud: Nonsuch Publishing, 2005.

Hudson, William Henry. 1892. *The Naturalist in La Plata*. New York: E. P. Dutton & Co., 1922.

Humboldt, Alexander von. *Views of Nature* [1808]. Translated by Mark W. Person. Edited by Stephen T. Jackson and Laura Dassow Walls. University of Chicago Press, 2014.

Humboldt, Alexander von. *Kleinere Schriften: Erster Band.* J. G. Cotta, 1853.

Humboldt, Alexander von. *Views of Nature, or, Contemplations on the Sublime Phenomena of Creation, with Scientific Illustrations* [1808]. Translated by E. C. Otté and Henry G. Bohn. Henry G. Bohn, 1850.

Humboldt, Alexander von. *Personal Narrative of Travels to the Equinoctial Regions of the New Continent, During the Years 1799–1804.* Translated by Helen Maria Williams. Longman, Hurst, Rees, Orme & Brown, 1821.

Latour, Bruno. *Science in Action: How to Follow Scientists and Engineers Through Society.* Harvard University Press, 1987.

Liedke, Heidi. *The Experience of Idling in Victorian Travel Texts, 1850–1901.* Palgrave Macmillan, 2018.

López, Francisco. "Blind Listening." In *The Book of Music and Nature*, edited by David Rothenberg and Marta Ulvaeus, 163–68. Wesleyan University Press, 2013.

López, Francisco. *La Selva.* Compact Disc. V2_Archief, 2008.

Lubrich, Oliver. "Humboldtian Landscapes." In *Environmental Aesthetics After Landscape*, edited by Jens Andermann, Lisa Blackmore, and Dayron Carrillo Morell, 73–109. Diaphanes, 2018.

Lubrich, Oliver, and Ottmar Ette. "Versuch über Humboldt." In *Alexander von Humboldt, Über einen Versuch den Gipfel des Chimborazu zu ersteigen* [1810–11], 7–76. Eichborn, 2006.

Morton, Timothy. *Dark Ecology: For a Logic of Future Coexistence.* Columbia University Press, 2016.

Olsen, Ida Marie. "Outlines of Ecological Consciousness in W. H. Hudson's Environmentalism." *English Literature in Transition, 1880–1920* 63, no. 2 (2020): 193–210.

Pratt, Mary Louise. *Imperial Eyes: Travel Writing and Transculturation.* Routledge, 1992.

Shrubsall, Dennis, and Pierre Coustillas. *Landscapes and Literati: Unpublished Letters of W. H. Hudson and George Gissing.* Russell, 1985.

Thompson, Marie. *Beyond Unwanted Sound: Noise, Affect, and Aesthetic Moralism.* Bloomsbury, 2017.

Tuan, Yi-Fu. *Topophilia. A Study of Environmental Perceptions.* Columbia University Press, 1990.

Velasco, Daniel. "Island Landscape: Following in Humboldt's Footsteps Through the Acoustic Spaces of the Tropics." *Leonardo Music Journal*, no. 10 (2000): 21–24.

CHAPTER 3

Archives of Extinction

Unproductive Bodies and Human/Nonhuman Expendability in the Argentine Desert

GISELA HEFFES

TRANSLATED BY ANDREA ROSENBERG

Después de la Europa, ¿hay otro mundo
cristiano civilizable y desierto que la América?

After Europe, is there any other uninhabited
and civilizable Christian world besides America?

— DOMINGO F. SARMIENTO, *Facundo*

The historiography of nineteenth-century Argentina often includes the long-standing controversy of the Alberdi-Sarmiento debate. This contention, of course, revolved around two different models for constructing the nation. Whereas Sarmiento advocated a centralized model with Buenos Aires as the capital, Alberdi argued for a federal model, governed by General Urquiza, and underscored the need to incorporate the Argentine provinces into the state-building project that followed Rosas's defeat in the Battle of Caseros in 1852. Thus are encapsulated two opposing positions that were historically complex stances, encompassing multiple contradictions and undergoing transformations over time.[1] At the center of the dispute was the question of what sort of nation would be constructed, as the Argentine historian Tulio Halperín Donghi put it in his book of the same name, "para el desierto argentino" (for the Argentine desert).[2] The notion of the desert is crucial as it signifies a sense of emptiness. In the collective imaginary, the desert is envisioned as a void sprinkled with only meager flora and fauna, almost entirely devoid of bioclimatic features or biotic diversity. It is a landscape whose perceived bareness harbors the potential for productivity, making it an apt, even ideal, space in which to erect the nation: a *terra nullius* vulnerable to being claimed by a colonizing state via the right of occupation.

In considering the marks that these nation-building projects left on the Argentine landscape over the *longue durée* of the nineteenth century, I explore two different aspects: first, the idea of the nation, along with the concomitant notions of economic progress and social transformation, which gradually gained ground, establishing elements of modernization; and second, the specific material qualities of the so-called Argentine desert, a tangible, living space, a biome inhabited by human and nonhuman creatures and a diverse array of plants, including forests of *caldén, chañar,* and *carquejilla* trees as well as (in the more arid western regions) different varieties of cactus. In addition to mammals, the more than three hundred species of birds include land and water birds such as the lesser rhea, the spotted nothura, and the maguari stork.[3] This essay takes an against-the-grain approach to conceptualizing the Argentine desert, one that stands in opposition to the "regime of emptiness," as Samia Henni calls it, and thus undermines the idea of legitimization that has accompanied its "transformation, manipulation, toxification, and destruction."[4] It is therefore important to restore these silenced histories and propose an alternative archive, one that reveals the palpable traces of extinction and reconsiders "modalities of and the relationships between and across desert zones," as well as those between deserts and urban and rural spaces.[5]

In *El discurso de la naturaleza: Ecología y política en América Latina* (1990), Fernando Mires describes the wages of progress: the indiscriminate slaughter of the Indians, who, not being "pueblos racionales" (rational peoples)—nor being rationalists, "en el sentido de rendir culto a la razón" (in the sense of venerating reason)—posed an obstacle to the logic of expansion and economic growth.[6] If they could not be integrated into a model of "desarrollo nacional" (national development), ipso facto they constituted an impediment whose elimination was defensible, legitimate, and justified. In nineteenth-century Argentina, these "obstacles" that Mires refers to consisted of a corporeal assemblage of *other* bodies, whether

Black, Indigenous, or gaucho. The desert is both a space that takes in these others and, at the same time, one in which bodies move around, territorial possession is disputed, and plans are destabilized. A site of desire and confrontation, the desert is an ambivalent and liminal space: the desert ever as abyss. Those who dwell in this supposedly uninhabited space are victims of an imaginary that incorporates not only every association related to the nation's future—that is, their own negation—but also, and even more so, the paradoxically irrational threat of potential destruction, death, and chaos. Once the Black inhabitants had disappeared and the Indigenous population had been exterminated, a different interest in the figure of the gaucho arose. Because, unlike the *other* others, the gaucho was credited with the possibility of productivity, some members of the *Rioplatense* intellectual elite were compelled to rescue this stigmatized figure and assimilate him in service of a rapid economic shift that was being implemented in post-Rosas Argentina—that is, after 1852.[7] The references to the gaucho's lack of productivity in *Facundo* (1845), a quality that ties him to a supposedly barren spatiality via a reading grounded in telluric determinism, gave way to a recovery of the paradigmatic but controversial figure, a recuperation that was, though slow, not free of ruthless violence. The gaucho exoticized, venerated, and abhorred by Sarmiento was "rescued," twenty-seven years later, by José Hernández in the first part of *Martín Fierro, La ida* (1871; *The Departure*). This rescue then became, in *La vuelta* (1879; *The Return*), moral condescension: having identified the danger of extinction, Hernández employs warnings and admonishments to urge the gaucho to give up his roaming ways and become a productive Christian laborer—and, through this conversion, to contribute to the new-fledged nation's economy. Thus, considering this regime of transformation, the present chapter examines the metamorphosis of the desert—and of the living beings that inhabit it—as that of a space whose changeover, paradoxically, gradually assembled a physical archive of human and non-human extinction. Whereas *Facundo* takes an ambivalent—even contradictory—stance on both the space and its inhabitants, José Hernández offers an insightful critique (in *La ida*) of the model of Sarmientan progress, formulated toward the end of *Facundo*, followed by a complacent subordination (in *La vuelta*) through which both the "matrero" (roaming) gaucho and the "ociosa" (idle) land physically disappear—that is to say, die out—and become instead instruments of economic change and thus, by extension, of social reconfiguration.[8] Because, as Halperín Donghi notes regarding the political platforms proposed by both Sarmiento and Alberdi, "[El] país necesita población; su vida económica necesita también protagonistas dispuestos de antemano a guiar su conducta en los modos que la nueva economía exige" (The nation requires settling; its economic life also requires protagonists prepared to guide its comportment in the modes that the new economy demands).[9]

Sarmiento: Lover of the Law

In the final chapter of *Facundo*, "Presente y porvenir" (Present and future), Sarmiento sets out the main contours of his ideas, already thoroughly developed over the course of the text.[10] His confidence in a prosperous future is apparent; his expectations rest on the new generation assembled around the Salón Literario in Buenos Aires. Despite their lack of experience, Sarmiento sees them as being the "precursores de un movimiento más fecundo en resultados" (precursors of a movement more fruitful in results).[11] He identifies them as having emerged from the "seno" (bosom) of this salon, describing them as "un grupo de cabezas inteligentes, que, asociándose secretamente, proponíase formar un carbonarismo que debía echar en toda la República las bases de una reacción civilizada contra el Gobierno bárbaro que había triunfado" (a group of intelligent leaders, who, meeting in secret, proposed the formation of a Carbonaria that would spread, throughout the Republic, the basis for a civilized reaction against the barbarous government that had triumphed).[12]

Simultaneously, he outlines some of the paradoxical events that occurred during Rosas's federal government, via a phrase that, in my view, sums up part of the ideology latent in the future Argentine president: "La *unión* es íntima" (Union is intimate).[13] Underlying that idea is the proposition for the kind of state that Sarmiento envisions for his nation's future. Additionally, while Sarmiento is noted for his exceptional clarity, his text reveals ongoing contradictions. Despite this, he operates from a distinct political and cultural platform, much like politicians launching their campaigns. Furthermore, Sarmiento is dedicated to an unrelenting project, to which he has devoted his entire life.

Asserting the intimate quality of unions means contemplating a number of elements that inevitably revolve around Sarmiento's central concepts: above all, the ineluctable need to lay the foundations for his project of erecting the state, a project already begun by that most "civilized" of men, Don Bernardino Rivadavia. Sarmiento describes him as la encarnación viva de ese espíritu poético, grandioso, que dominaba la sociedad entera. Rivadavia, pues, continuaba la obra de Las Heras en el ancho molde en que debía vaciarse un grande Estado americano, una República. Traía sabios europeos para la prensa y las cátedras, *colonias para los desiertos*, naves para los ríos, interés y libertad para todas las creencias, crédito y Banco Nacional para impulsar la industria; todas las grandes teorías sociales de la época, para moldear su gobierno; la Europa, en fin, *a vaciarla de golpe en América*, y realizar en diez años la obra que antes necesitara el transcurso de siglos.

the living incarnation of the poetic, grandiose spirit that dominated the entire society. Rivadavia, then, continued the work of Las Heras on the large mold into which a great American state, a republic, would be poured. He brought in learned Europeans for the press and the universities, *settlement colonies for the deserts*, ships for the rivers, concern and freedom for all beliefs, credit and the Banco Nacional to stimulate industry, all the great social theories of the era for modeling his government; in short, he brought in Europe, so as *to pour it all at once into America*, and to realize in ten years the work that before would have required the passing of centuries.[14]

Pouring Europe into America "all at once" implies a priori conceiving of America—and especially Argentina—as empty. It denies not only human but also nonhuman existences. It proposes a total transfer to fill the "void" with living organisms foreign to the local landscape and cram it with invasive species. To this end, Sarmiento suggests that all of the fundamental elements of civilization fuse together and flow toward a common destination, giving them legal form and providing them with materiality and legitimacy, crafting a need that can unify them and through which their very substance can be stamped with an original identity: "porque las fórmulas legales son el culto exterior que rinde a sus ídolos, la Constitución, las garantías individuales" (because legal formulas are the external worship he pays to his idols: the Constitution, individual rights).[15] Rivadavia's religion, he says, is "el porvenir de la República, cuya imagen colosal, indefinible, pero grandiosa y sublime, se le aparece a todas horas cubierta con el manto de las pasadas glorias y no le deja ocuparse de los hechos que presencia" (the future of the Republic, whose image—colossal, indefinable, but grand and noble—appears to him at all times covered with the mantle of past glories and doesn't allow him to think about the events he is witnessing).[16] Paradoxically, however, the desired "union" was finally achieved, Sarmiento claims, as a result of Rosas's despotism. Through his persecutions, Rosas managed to disappear the *montoneras* from the countryside, give the gauchos a "more civilized" aura, and impel a large number of Buenos Aires residents to settle rural areas, bringing their civic spirit with them. Accordingly, he writes, "La unidad de la República se realiza a fuerza de negarla; y desde que todos dicen federación, claro está que hay unidad" (The unity of the Republic is realized by dint of denying it, and since everyone is saying "federation," of course there is unity).[17] In this same section, Sarmiento underscores the social changes that took place thanks to the governance of the most tyrannical, systematic, cold, and calculating of all the caudillos.[18] Despite his emphasis on (and interest in) some form of social equity, in the Sarmientan binary, not all individuals are positioned to take on the civic commitment that is so eagerly—even desperately—sought. This comports with the fact that within the vaunted "unity" there are those whom the "enlightened" cause of "civilization" does not favor: of the Black "race," Sarmiento writes, "Felizmente, las continuas guerras han exterminado ya la parte masculina de esta población, que encontraba su patria y su manera de gobernar en el amo a quien servía" (Happily, continual wars have now exterminated the masculine part of this population, who found their own homeland and its form of government in the master whom they served).[19] Within an archive of extinction, the Black "race" points to not only a justification for its disappearance but also the joyous celebration of a biological continuity that has been cut short. According to the historian James Scobie, the "disappearance" of Argentina's Black population was a mystery to demographers given that

> by 1810 Negroes and mulattoes constituted a significant element of the urban population and probably numbered 60,000. By 1895 this total had decreased to 5,000, and, although this figure is taken as a current estimate of the Negro element in Argentina, it is probably too high. The struggles for independence and the subsequent civil wars, as well as tuberculosis and other diseases, carried off an inordinately high percentage of Negroes and mulattoes, but even more effective was their gradual absorption into the dominant creole-mestizo population.[20]

The "raza salvaje" (savage race) brought from Africa is prevented from reproducing and, as a result, loses the continuity necessary to ensure its own racial, cultural, and social perpetuation.[21] At the same time, the other "savages" who dwell in the desert, a liminal border zone undergirded by pure barbarism, must, without exception, disappear. The gaucho remains. As such, it is worth wondering why Sarmiento is so interested in rescuing both the figure of the gaucho (in descriptive, almost pictorial terms) and his physical self, his factual presence. What *a posteriori* project does he envisage based on this conflictive presence, and what purpose does it serve? In a reverse reading—that is to say, one that engages in a "negative" enumeration of the qualities that Sarmiento attributes to the gaucho—we can point to one concept that is noticeably contradictory (though still ultimately coherent): Sarmiento's notion of progress and national *unity* is inexorably tied to the gaucho being perpetually occupied—as Spanish etymology suggests, rejecting idleness (*neg-ocio*), busy with his business. And if business implies the absolute rejection of the much-coveted leisure time, it is because free time brings with it—also inexorably—the taint of vice, crime, and, ultimately, *unproductivity*. If the gauchos persist indefinitely in this state of idleness—one that, as previously noted, mirrors the condition of the land they inhabit—there will be no advancement in the countryside, no economic growth, and no possibility of social reorganization. Consequently, the territorial framework that Sarmiento predicts in his political map, where he aims to realize his vision for the Argentine Republic, will remain unattainable. Scobie's description is eloquent:

> The estancia headquarters was not the palatial weekend or summer retreat which it would become at the end of the nineteenth century. Those eccentrics who managed their own properties shared the existence of their gauchos. Their diet was meat and mate. They wore simple clothing suited to wind, rain, dust, and equestrian activity. Their home was a mud hut, or rancho, furnished with a cot made of hides, some ox skulls for chairs, and a few pegs from which to hang the heavy silver spurs and ornate bridles. Twenty or thirty miles separated neighbors, and the charred remains of more than one rancho bore eloquent testimony to raids by marauding Indians. Little wonder then if the average estanciero satisfied himself with an infrequent visit to his holdings or an occasional accounting by his foreman.[22]

It is therefore vital to initiate a transformation process that reverses the reality that enfolds not only the gauchos but also the ranchers as well as the economic potential of the space they inhabit. In other words, organizing the desert; organizing the pampas. Transforming that untapped power into manpower; employing it in an enterprise of economic production. Heading into the future: a future that can be reached only hand in hand with progress and alongside civilized peoples. Furthermore, Sarmiento is not only concerned but irritated that this kind of occupation or daily pseudo-activity, which routinely sustains thousands of gauchos, gives them pleasure, joy, jubilation: the gaucho "no trabaja; el alimento y el vestido lo encuentra preparado en su casa; uno y otro se lo proporcionan sus ganados, si es propietario; la casa del patrón o pariente, si nada posee" (does not work; he finds food and clothing at hand in his home. Both of these are provided by his livestock, if he is a proprietor, or the house of his employer or relatives, if he owns nothing).[23] On the other hand, "las atenciones que el ganado exige se reducen a correrías y partidas de placer. La hierra, que es como la vendimia de los agricultores, es una fiesta cuya llegada se recibe con transportes de júbilo: allí es el punto de reunión de todos los hombres de veinte leguas a la redonda; allí, la ostentación de la increíble destreza en el lazo" (the attention the livestock require boils down to excursions and pleasurable games. The branding, which is like the grape harvest for farmers, is a celebration whose arrival is greeted with transports of joy: this is the place where all the men within a twenty-league radius meet, where they show off incredible skill with the lasso).[24] In the eyes of the bourgeois intellectual from San Juan (or, as David Viñas refers to him, the "bourgeois conquistador"), whose economic model is based on "accumulation," the gauchos—and their occupation—themselves constitute an entity that operates outside the law.[25] Lacking external laws and existing outside of any institutional organization, they therefore also lack morality. If the state behaves in accordance with the prevailing moral order in society, the most pressing issue is to modify the social components that make up society's fundamental substrate, suggesting a need to organize the state in keeping with the most elementary characteristics. Thus, a barbaric, savage society produced Rosas—but Buenos Aires society produced the Unitarios. Sarmiento offers an alternative history in which he attempts to show the reader what the Argentine Republic might have been without the ten years of tyranny, and what the "new government" of the future will be like, once the tyrant has been overthrown: "se rodeará de todos los grandes hombres que posee la República, y que hoy andan desparramados por toda la tierra, y con el concurso de todas las luces de todos hará el bien de todos en general. La inteligencia, el talento y el saber serán llamados, de nuevo, a dirigir los destinos públicos, como en todos los países civilizados" (the Republic will surround itself with all the great men it possesses, who today are spread all over the world, and with the combination of all those lights will do good for everyone in general. Intelligence, talent, and knowledge will be called upon once again to direct the public destiny, as in all civilized countries).[26]

It is worth keeping in mind *Facundo*'s narrative complexity: beyond the way it plays with established binaries to differentiate (in an exaggeratedly Manichean fashion) the elements associated with the concepts of civilization and barbarism, an interaction and ongoing permeability between the two spheres nonetheless prevails. And although this dynamic highlights the enormous challenge of employing categories in a superficial, even hasty manner, it nevertheless still sets out the most important foundations for the project inherent to its ideology. Of course, the presumed correspondence between society and a singular moral system runs the risk of delegating to the state a coercive power with totalitarian qualities,[27] which would, ironically, coincide with the politics implemented by Rosas: in this case, that of a state whose totalitarian morality was imposed via the Mazorcas, the *montoneras*, and the "Lomo Negro" Federales, and through which the elements of an unquestionable social and governmental framework took root.

But Sarmiento has no intention of expanding his state project to incorporate cultural complexity like that exhibited by a society made up of people of multiple origins, "razas" (races), and a variety of religious practices. Instead, he proposes replacing one carefully delimited state with another, equally closed off: the notion is based on replacing the totalitarian Rosista state, rooted in "barbarism," with one that is "civilized"—we might even say Pazifista, given the end of *Facundo*, which enthusiastically endorses the wait for another Restorer, General Paz—in other words, one whose foundations are erected atop the model of European "civilization." It is, in some sense, totalitarian, given that it accepts no other model but the European model, just as it accepts no other population besides that of the old continent (the success of the nation's progress will depend on the kind of immigration it receives). Sarmiento, therefore, outlines a tightly cropped society, with a large array of others being left outside its boundaries. And in this spatial demarcation—human and nonhuman alike—the image of the barbed-wire fence embodies both the idea of the boundary and that of economic productivity: division based on private property but, fundamentally, as an emblematic image of the controlling state.

In Praise of Unproductivity

José Hernández's *Martín Fierro*, specifically *La ida* (1871), is, to some degree, a tale of resistance, rebellion, and defiance. It differs from the second part (*La vuelta*, 1879) not only in its action-packed plot but also in its defense of rural people in the face of the self-importance and exploitativeness of the city—that is to say, of Buenos Aires's educated elites—and thus stands in opposition to the Sarmientan binary of civilization versus barbarism.[28] Most of these problems had already been explored by Hernández in his extensive journalistic work. Faced with dizzyingly rapid industrialization, a republic crisscrossed by railroad tracks and barbed-wire fences and increasingly inundated with immigrants, the pampas became an ever more hostile space for the gauchos, who were victims of the infamous "malones" (surprise Indian raids) and, at the same time, served as cannon fodder for the nation's wars. *Martín Fierro* embodies resistance to the forces of oppression and exploitation legitimized by law and progress—forces that, for Sarmiento, represent the foundations of civilization. More to the point, the text constitutes an effort to save a species on the brink of extinction. To this end, Hernández questions the ontological classification of the gaucho, who is deemed fundamentally criminal ("el ser gaucho es un delito" [to be a gaucho is a crime]).[29] In the "Carta Aclaratoria" (Explanatory Letter) addressed to José Zoilo Miguens, he announces his desire to "retratar, en fin, lo más fielmente que me fuera posible, con todas sus especialidades propias, ese tipo original de nuestras pampas, tan poco conocido por lo mismo que es difícil estudiarlo, tan erróneamente juzgado muchas veces, y que, *al paso que avanzan las conquistas de la civilización, va perdiéndose casi por completo*" (in sum, portray as faithfully as possible, with all of its particularities, that original type of our pampas, so little known that it is difficult to study it, so often misjudged, and that, *as the conquests of civilization advance, is gradually being almost entirely lost*).[30] Starting in the third canto, we see the inequity, harm, and injustice inflicted by the "autoridad" (authorities) on the gauchos, emphasizing the corruption of the judges and depicting the Indians—not the gauchos—as the "bad guys." Similarly, the text refers to the presence of the foreigner, or "gringo," a quarrel with whom sets off a questioning of his role on the frontier: "Yo no sé por qué el gobierno / nos manda aquí a la frontera / gringada que ni siquiera / se sabe atracar a un pingo" (I don't know why the Government sends us, out here to the frontier, / these gringos that don't even know how to handle a horse).[31] And it is made clear, in the first part of *Martín Fierro*, that the extermination of the Indian is more than justified: "pues donde dentra / roba y mata cuanto encuentra / y quema las poblaciones" (where they break in / they'll steal and kill all they come across and burn down the settlements).[32]

The first-person voice of the gaucho is a construct, as Josefina Ludmer points out in her now classic *El género gauchesco* (1988).[33] Hernández employs this injustice-denouncing voice to express concern and highlight the gauchos' woeful status. As a member of the Partido Federal (even if he did not support the Rosas dictatorship), Hernández held a political stance that differed from Sarmiento's, and, by extension, with regard to the "problem" of the countryside, also from that of most educated elites (who, like Sarmiento, viewed the agricultural model and immigration as two elements fundamental to attaining the much-vaunted civilization and thus giving shape to the project of the nation). In a letter to the editors of the eighth edition of *Martín Fierro*, Hernández writes that

> un país cuya riqueza tenga por base la ganadería, como la provincia de Buenos Aires y las demás del litoral argentino y oriental, puede no obstante, ser tan respetable y tan civilizado como el que es rico por agricultura o el que lo es por sus abundantes minas o por la perfección de sus fábricas. La ganadería puede constituir la principal y más abundante fuente de riqueza de una nación.

> a country whose wealth is based on livestock, such as the province of Buenos Aires and the other provinces of the Argentine and Uruguayan coasts, can nonetheless be so respectable and civilized as one that is wealthy because of agriculture or one that is so because of its plenteous mines or the excellence of its factories. Livestock can constitute a nation's primary and most abundant source of wealth.[34]

Far from believing, as Sarmiento did, that barbarism could be addressed only via the firm imposition (and importation) of a "civilizing" model, whose maximum expression in the pampas was agriculture, Hernández saw the economic potential of cattle-raising. In *La ida*, the gaucho is a marginal figure who is presented as "duro" (hard); his greatest glory is living free, he has no home, he dwells under the open sky, he fights if necessary, and because he is (unjustly) persecuted, he becomes a bandit.[35] The "campo" (plain), that vast and fearsome expanse, lacks, as Borges has noted, description: in *Martín Fierro* there is no landscape, as Hernández "presupone deliberadamente la pampa y los hábitos diarios de la pampa, sin detallarlos nunca—omisión verosímil en un gaucho, que habla para otros gauchos" (deliberately presumes [knowledge of] the pampas and their daily customs without ever detailing them—a believable omission in a gaucho, who speaks to other gauchos).[36]

Beyond the narrative strategy, evoking this space as a desert landscape invites a conceptualization of its aesthetic potentiality as a space of absence. Yet this region contained rivers, pastureland, *caldén* groves, human and nonhuman beings. Although descriptions of the landscape are scarce in La Ida, certain allusions evoke a kind of community of the living. It is to the "plain" that Martín Fierro turns, "all alone" and "wild," because not only does the open air offer him refuge but he also establishes there an alliance that protects and guides him: "Y al campo me iba solito / más matrero que el venao" (And I'd go off on the plain all alone, wild as a deer), and it is also on the "plain" that, "sin punto ni rumbo fijo / en aquella inmensidá, / entre tanta escuridá / anda el gaucho como duende / allí jamás lo sorpriende / dormido, la autoridá" (without an aim or a fixed course in that immensity, / with that great darkness round him, a gaucho roams like a ghost— / out there, the authorities will never catch him asleep).[37] At the same time, "sin más amparo que el cielo" (with no help except from heaven), for the gaucho the stars are allies in moments of seeking comfort and guidance: "Ansí me hallaba una noche / contemplando las estrellas, / que le parecen más bellas / cuanto uno es más desgraciao / y que Dios las haiga criao / para consolarse en ellas. // Les tiene el hombre cariño / y siempre con alegría / ve salir las Tres Marías, / que, si llueve, cuanto escampa / las estrellas son la guía / que el gaucho tiene en la pampa" (And so one night, I was out there gazing at the stars, / which it seems are more beautiful the more unhappy you are, / and that God must have created them for us to find comfort there. // A man feels love for them, and it's always with joy / that he sees the Three *Marías* coming out—because when there's been rain, / as soon as it clears, on the pampa, the stars are a gaucho's guide).[38]

But the ecosystem of the desert (especially the *pampa seca* to the west, the semi-arid plains that extend into Patagonia), to which the gaucho Martín Fierro and his friend Cruz plan to flee near the end of *La ida*, offers a diverse panoply of plant life and wild animals. The poem highlights the region's diversity, describing it as packed with a large variety of species: "De hambre no pereceremos, / pues según otros me han dicho / en los campos se hallan bichos / de los que uno necesita . . . / gamas, matacos, mulitas, / avestruces y quirquinchos" (We won't die of hunger, as according to what I've been told, / in the wild lands there are animals, all the kinds you need— / wild does, deer, mulitas, armadillos and ostriches).[39] Nor is it strictly "desert": "Tampoco a la sé le temo, / yo la aguanto muy contento, / busco agua olfa-tiando al viento, / y dende que no soy manco / ande hay duraznillo blanco / cavo y la saco al momento" (I'm not afraid of thirst, either, I can bear it quite cheerfully— / I can find water sniffing the wind, and while I'm still sound of limb / I can dig and reach it right away anywhere there's a white-peach tree).[40] We know that the supposed alliance with the Indians posits a soli-darity that will be undermined in the second part (La vuelta). But it is in La ida where the idealization of the flight—that crossing of borders—heralds a future in which unproductivity is celebrated: "Allá no hay que trabajar, / vive uno como un señor; / de cuando en cuando un malón, / y si de él sale con vida / lo pasa echao panza arriba / mirando dar güelta el sol. // Y ya que a juerza de golpes / la suerte nos dejó aflús, / puede que allá véamos luz / y se acaben nuestras penas. / Todas las tierras son güenas: / vámonos, amigo Cruz" (Over there, there's no need to work, you live like a lord— / going on a raid from time to time, and if you get out from that alive / you live lying belly-up watching / the sun go round. // And now that Fate has beaten us and left us high and dry, / maybe we'll see light, over there, and our sorrows come to an end. / Any land will do for us—let's be going, Cruz my friend).[41]

The celebration of not working and unproductivity, within the logic of capitalism, equates the "infertility" of the land with the "sterility" of the gaucho. In contrast, Martín Fierro rejects a notion of national identity rooted in the pursuit of economic progress. Furthermore, it challenges the concept of "imaginación territorial" (territorial imagination), which links the desert to terra nullius.[42] And it dismantles the ideological framework that systemat-ically denies the existence of living organisms (human and nonhuman) co-habiting in the pampa seca. In La vuelta, the living quality of the landscape grows darker. Whereas La ida traces the image of "that original type of our pampas," the "deserter" whom the military views as a threat to the ideol-ogy of the time, La vuelta creates an inverted image: now we are given the paradigm of the "regenerado" (regenerate), whose rehabilitation emerges as a voice that calls on a series of civilized elements tied to the project of establishing the nation. As such, whereas La ida is "barbaric," La vuelta is "civilized"—always in quotation marks. The consummation of the offen-sive tactic in the Argentine desert—that is, the brutal expulsion and subse-quent annihilation, imprisonment, and even enslavement of the Indians—

together with the resulting usurpation, incorporation, and distribution of their lands, put an end to a long series of wars between caudillos and Liberales, and coincided (not fortuitously) with the technological emergence of a modernity whose great emblems would be the invention of refrigeration and the laying of the railroads, making way for beef exportation and Argentina's entry into the international market. The unproductiveness celebrated in *La ida* is resignified in *La vuelta*, and this regeneration is a radical one, now conveying the negative experience of the past and the virtues offered by this transformation process. Martín Fierro becomes a father who gives advice and presents himself as a measured person. Whereas in *La ida* crime was justified by necessity and perseverance, now the gaucho must be restrained, balanced. In a sort of repertory of virtues, Hernández offers a catalog that includes being devoted to family and closely connected to siblings, being diligent and prudent, respecting the elderly, being tender like the "cigüeña" (stork), being obedient—given that obedience is synonymous with goodness—giving up vices and having good judgment, not stealing, not fighting, not killing anyone, not getting drunk, and respecting women. But, above all, "el trabajar es la ley" (to work is the law).[43]

As the landscape inevitably changes, Hernández calls for the transformation of the gaucho. Moreover, he presents a series of warnings and recommendations for assimilating the gaucho into human labor. Whereas the Indian has been annihilated, the gaucho will be assimilated. The suggestions set out in *La vuelta* add the gauchos to the growing archive of extinction.

The Barbarity of Civilization

To achieve the aforementioned union that Sarmiento so vehemently sought—that is, to *unify the nation*—we must draw lines and grid the map. We must test out a geometry that combines the railways, the implementation of barbed-wire fences, and the incorporation of new settlers, establishing them against the longstanding native inhabitants, living organisms who survived the Conquista del Desierto led by Julio Argentino Roca in 1879. Anything not wiped out by slaughter or, as David Viñas called it in *Indios, ejército y fronteras*, "genocidio" (genocide). Here Viñas wonders,

Si en otros países de América Latina la "voz de los indios vencidos" ha sido puesta en evidencia, ¿por qué no en la Argentina? ¿La Argentina no tiene nada que ver con los indios? ¿Y con las indias? ¿O nada que ver con América

Latina? Y sigo preguntando: ¿No hubo vencidos? ¿No hubo violadas? ¿O no hubo indias ni indios? ¿O los indios fueron conquistados por las exhortaciones piadosas de la civilización liberal-burguesa que los convenció para que se sometieran e integraran en paz? ¿Y qué significa "integrarse"? Pero, me animo a insistir: ¿por qué no se habla de los indios en la Argentina? ¿Y de su sexo? ¿Qué implica que se los desplace hacia la franja de la etnología, del folclore o, más lastimosamente, a la del turismo o de las secciones periodísticas de *faits divers*? Por todo eso me empecino en preguntar: ¿no tenían voz los indios? ¿O su sexo era una enfermedad? ¿Y la enfermedad su silencio? Se trataría, paradójicamente, ¿del discurso del silencio? O, quizá, los indios ¿fueron los desaparecidos de 1879? Todos esos interrogantes, especialmente ahora, necesito aclararlos. Lo intentaré, trataré de hacerlo. Dado que, francamente, no me convence la versión que me ofrece el circuito liberal de 1879 hacia acá.

If in other Latin American countries the "voice of the vanquished Indians" has been highlighted, why not in Argentina? Does Argentina have nothing to do with [male] Indians? And with female Indians? Or nothing to do with Latin America? And I ask too: Were none vanquished? Were none raped? Were there no Indians, either male or female? Or were the Indians instead conquered via the pious exhortations of liberal-bourgeois civilization, which persuaded them to submit and peacefully integrate? And what does "integrate" mean? But, I dare to insist: Why are Indians not talked about in Argentina? What about their sex? What does it mean that they are shoved to the margins of ethnology and folklore, and, more regrettably, to tourism and lurid tabloid stories? Because of all this, I insist on asking: Did the Indians have no voice? Or was their sex a disease? And that disease their silence? Could it, paradoxically, be the speech of silence? Or, perhaps, were the Indians the disappeared of 1879? I must shed light on these questions, especially now. I will try; I will make an attempt. Since, frankly, I am not convinced by the version offered by the liberal circuit from 1879 until now.[44]

Esteban Echeverría opens his renowned poem *La cautiva* (1840) by describing the desert, which, "inconmensurable, abierto / y misterioso a sus pies / se extiende, triste el semblante, / solitario y taciturno / como el mar, cuando un instante / el crepúsculo nocturno / pone rienda a su altivez" (immeasurable, open, / and mysterious at their feet / spreads, gloomy of visage, / solitary and taciturn / as the sea, when in a moment / the nocturnal crepuscule / reins in their haughtiness).[45] Echeverría later notes that the dazzling expanse is inhabited by a "bando de salvajes" (band of savages) who, "atronando todo el campo convecino" (deafening all the neighboring land), like a "torbellino" (whirlwind), "hiende el espacio veloz" (swiftly slashes the space).[46] The text reinforces the imaginary of a desert devoid of life—or, rather, one that is home to "barbarians," who, having been excluded, are automatically reclassed as inhuman. This maneuver precedes the political justification of what was to take place twenty-nine years later. Not only do desert territories comprise approximately one-third of the land mass on Earth and are home to human and nonhuman, biological and microbiological lives, but they also sustain various forms of existence: sedentary, nomadic; animal, vegetable, and mineral. As Samia Henni notes, the repeated insistence on defining the desert as an inert space, configured within a "regime of emptiness," has no other objective than to emphasize, in service of a logic of exploitation and industrialization, the need to occupy or fill these territories through acquisition, extraction, mining, production, and accumulation.[47] These phenomena, tied to implicit or explicit forms of colonization and toxicity, result in the racialization, alteration, injury, and destruction of the living environments, natural and/or constructed, that the desert contains.[48]

If Sarmiento's aim was to pour Europe into America, Alberdi's famous dictum asserts that "gobernar es poblar" (to govern is to populate):

> Gobernar es poblar; pero poblar es un arte, una ciencia, el arte, la rama más importante de la ciencia del gobierno, que es la economía política, es decir, la economía discreta, juiciosa, que no comete la impolítica de confundir la población mala con la buena, despoblando en vez de poblar; porque envenenar un país física y moralmente, es despoblarlo y hacerlo retroceder más atrás de la barbarie. El gobierno tiene un poder eficaz de selección en materia de población. No con reglamentos y prohibiciones de que se burla la naturaleza de las cosas; sino con diques, con obras, digámoslo así, como las que cambian las corrientes naturales de los ríos y de las aguas más libres.

> To govern is to populate; but populating is an art, a science, art, the most important branch of the science of governing, which is political economy—that is, the discreet and wise economy, which does not impolitically confuse bad population with good, depopulating rather than populating;

because to poison a country physically and morally is to depopulate it and cause it to regress even beyond barbarism. The government has an effective selection power when it comes to population. Not with rules and prohibitions flouted by the nature of things; but with dams, with works, we might say, such as those that change the natural currents of the rivers and of the freest waters.[49]

Here technology is evoked as serving the government to fill the void with "good" population; the "bad" population is toxic and therefore poisons a society. And when it contaminates, it depopulates. It regresses or, in other words, undoes the forward movement that guides progress and retreats into "barbarism." Alberdi clearly understands that the interference of the technological advances on which modernization processes rest will "naturally" bring with it works constructed by men, which would have a palpable effect on the nearby populations (creating the conditions, again "natural," for selection among humans and nonhumans). The push to arrange the *emptied* space, in tandem with its human and nonhuman bodies, through the technologies of power is well known. What is surprising about Alberdi, however, is that alongside his promotion of a eugenicist praxis (*Peregrinaciones* was written in 1871, the same year as Hernández's *La ida*), he also recognizes, with humor or cynicism—or both—that civilization cannot exist without barbarism:

Seamos justos. ¿Qué es nuestra civilización sino la barbarie regularizada? ni ¿qué es la barbarie sino la materia primera de que está fabricada nuestra civilización? Civilizado o bárbaro, el hombre vive del robo: toda la diferencia está en la forma del pillaje. Desnudo y desarmado, el hombre nace conquistador y usurpador por derecho. Examinad su persona de pies a cabeza: todo lo que viste es ajeno, y lo tiene contra la voluntad de su dueño. No dirá él que el ternero ha consentido gustoso en que le saquen el cuero de que está formado el calzado que visten sus pies; ni que el cabrito le ha regalado su propio pellejo para que vista sus manos con el guante que las abriga. La lana de que está hecho el vestido que cubre su cuerpo, pertenece a los carneros que han quedado desnudos, a la intemperie, para que el hombre cubra su desnudez. La seda de su corbata y de su sombrero ha sido el traje de gusanos, que han quedado desnudos para que el hombre se adorne con su precioso producto. ¿De qué se alimenta el hombre más civilizado y más cristiano? de cadáveres de animales, que lejos de dañarle, han sido a menudo sus mejores servidores y amigos: las gallinas y los pichones por ejemplo. Su mesa diaria es un anfiteatro anatómico; una carnicería hecha a sangre fría; un montón de cadáveres o de vivientes que han sido muertos, para que el hombre viva, y viva bien, y lo mejor posible. ¿Qué es la cama en que duerme? Lana y pluma, que han dejado desnudos o sin vida a sus dueños naturales.⁵⁰

Let us be fair. What is our civilization if not regularized barbarism? And what is barbarism if not the raw material from which our civilization is fashioned? Civilized or barbarian, man lives by theft; the sole difference is in the form of looting. Naked and unarmed, man is born a conqueror and usurper by right. Study his person from head to toe: everything he wears belongs to something else, and he possesses it against its owner's will. He will not claim that the calf has gladly consented to the removal of the leather from which the footwear in which he is shod is made; nor that the kid has given him its own hide so that he can dress his hands with the glove that shields them. The wool from which the suit that covers his body is made belongs to the rams that have been left naked, exposed, so that man can cover his nakedness. The silk of his tie and hat were once the garb of worms, which have been left naked so that man can adorn himself with their precious product. What does the most civilized and Christian man feed on? On the corpses of animals that, far from harming him, have often been his finest servants and friends: chickens and pigeons, for example. His daily table is an anatomical dissection hall; a cold-blooded butcher shop; a heap of corpses or living creatures that have been killed so that man lives, and lives well, and as well as possible. What is the bed in which he sleeps? Wool and feathers, which have left their natural owners naked or lifeless.

This recognition, which might initially appear as a moment of clarity, ultimately reveals a problematic cynicism. Alberdi suggests that we are all complicit in and accountable for barbarism. Following this reasoning, he legitimizes a range of human behaviors that, in their everyday manifestation, place us all on equal footing. In other words, Alberdi posits that we are all guilty—albeit to varying degrees—of barbaric acts, from the government wielding its "effective power of selection" to the most innocuous individual resting in their bed. As such, if each human being is equally responsible for the ongoing extractive and destructive practices, and for systematic consumption, despite the scale, then the government and its "effective selection power" are blameless. The message is clear: there is no distinction between actions taken at the government level and those practiced by individuals (all of us participate in a chain of extractions). This equalizing rhetoric thus levels out responsibilities and erases the impact that some decisions have over others. And shuts off not only any possibility of accountability but also any form, or possibility, of dialogue.

Coda: Shrunken Landscapes, Expanded Imaginaries

The shrinking of the pampas landscape thanks to the *barbarism of civilization* is a fundamental fact that calls for reassessing not only the debates over the civilization-barbarism binary but also, and especially, the notion of national identity. What does it mean to make use of a presumably empty space and develop a geographical imaginary that considers it devoid of any life forms? What does it mean to resort to external populations to "fill" or, rather, to "populate" those territories that are presumed to be empty? It means classifying the existent living organisms that inhabit these lands as expendable. This aspect, which David Pellow studied within the framework of his proposal for a critical environmental justice, is related to the effort to tackle "the largely unexamined question of the expendability of human and nonhuman populations facing socioecological threats from states, industries, and other political economic forces."[51] Populating or "filling" supposedly empty spaces with subjects from far-flung places all over the planet constitutes an immediate violence exerted against the land and its native residents—erasure as a requisite for "unification." Nevertheless, as is common knowledge, the vast majority of immigrants ended up settling in urban areas. Lettered reason's dream of transforming the Argentine desert into pure "civilization" remained incomplete. But positing an urban-rural divide, like the civilization-barbarism binary, is not merely superficial but also reductive. As Alberdi suggests in *Bases* (1852), the countryside is the city's condition of possibility, and vice versa:

> No hay otra división del hombre americano. La división en hombres de la ciudad y hombres de las campañas es falsa, no existe. . . . Rosas no ha dominado con gauchos sino con la ciudad. Los principales unitarios fueron hombres del campo tales como Martín Rodríguez, los Ramos, los Miguens, los Días Vélez; por el contrario, los hombres de Rosas, los Anchorena, los Me-

drano, los Dorrego, los Arana, fueron educados en las ciudades. La mazorca
no se componía de gauchos.

There is no other divide in the American man. The divide between city
men and country men is false; it does not exist. . . . Rosas has triumphed
not with gauchos but with the city. The foremost Unitarios were men of
the countryside such as Martín Rodríguez, the Ramoses, the Miguenses,
the Días Vélezes, whereas Rosas's men—the Anchorenas, the Medranos,
the Dorregos, the Aranas—were brought up in the cities. The Mazorca
was not made up of gauchos.[52]

The divide—the definitive one, in Alberdi's view—is between "European" men and "savage" ones, between an inclusive second person and an
excluded otherness. The gaucho's unproductive body, and the way it of-
fends Sarmiento, is, for Alberdi—as it is in Hernández's *La vuelta*—a productive force capable of fostering an economy based on the ideal of progress.
As Andrea Cobas Carral suggests, Sarmiento "confunde el origen de todos
los problemas políticos y civiles argentinos al adjudicárselos al influjo de la
Pampa" (jumbles the source of all of Argentina's political and civil problems when he attributes them to the influence of the Pampas); Alberdi, on
the other hand, "tuerce la división sarmientina entre campo y ciudad, par-
tición que ubica el atraso en el primero y los signos del progreso en la se-
gunda" (twists the Sarmientan divide between country and city, a division
that situates backwardness in the former and signs of progress in the latter), since, for Alberdi, "en toda civilización, las ciudades cobran impulso a
partir de las riquezas y el trabajo producidos en las zonas rurales" (in every
civilization, the cities are powered by the wealth and labor produced in the
rural areas).[53] After all, it is in the riches of the countryside that the power
of the cities resides.[54] By thinking in terms of reciprocity and correlation,
this approach reconfigures an isolating notion of the landscape—namely,
the Argentine desert—destabilizing any categorization that seeks to untan-
gle it from seemingly "external" elements and influences. Such exteriority
does not exist, as ecosystems, terrestrial and ideological alike, are complex
tapestries that go far beyond simplistic, Manichean binaries.

Nevertheless, there is little scholarship analyzing the configuration of
the living organisms typical of the *pampa seca* and/or the Argentine desert. According to Denis Cosgrove, landscapes serve as discourses through
which historically identifiable social groups have defined both themselves

and their relationship with the land and with other human groups, and these discourses are closely tied, epistemically and technically, with specific manners of seeing.[55] Cosgrove highlights not just the gaze that defines spaces but also the politics that uses landscapes to measure and regulate the social body. The Argentine desert, geographically articulated through textual and visual imaginaries, is inscribed within an abundant array of narratives. Jorge Luis Borges, Benito Lynch, Ezequiel Martínez Estrada, Osvaldo Lamborghini, Roberto Bolaño, Pablo Katchadjian, and Oscar Fariña as well as Leopoldo Torre Nilsson, Fernando Solanas, and Humberto Cairo are just a few examples. These rewritings, however, omit any explicit reference to the relationship between the landscape and the geological changes that shape its physical characteristics. Nor do they make any effort to grapple with the impact of the ecological catastrophe. Contrasting with this imaginary configuration of Argentina's *pampa seca*—the desert—the work of Gabriela Cabezón Cámara stands out. Her novel *Las aventuras de la China Iron* (2017; *The Adventures of China Iron*) not only seeks to challenge traditional genealogies but also proposes aesthetic, historical, and political perspectives that reimagine the desert landscape as inhabited by a diverse array of beings. This includes not just gauchos and *Indios*, but also women and various forms of life.[56] These existences, silenced from official histories, narrate stories of crossing boundaries in various ways. It is not just the physical space and landscapes that change; the nature of writing, the locations of power that define or have defined places for expression, and the characters involved in these narratives are also transformed. This transformation of the landscape reflects therefore a dynamism that seeks to implement new ways of understanding these archives of extinction. The fossils buried deep in the earth not only reemerge as once-silenced voices, but also are manifested as recollection, collective memory, or remnants of a past that was gradually assimilated until disappearing in the most absolute sense. But it is not merely disappearance; it is forced disappearance, territorial appropriation and expulsion. This territorial vastness reminds us that the archives of extinction are repositories assembled starting from a place of systematic violence. A systematic violence backed by the state.

In addition to rewriting, *Las aventuras* presents a new phenomenology and invites us to establish relational connections between humans and nonhumans, living and inert organisms, in order to create a space of connectivity that includes and joins together all the manifestations of life that dwell on the planet. In *Las aventuras*, Gabriela Cabezón Cámara rewrites *Martín Fierro*, upends it, dismantles it to propose, as Borges did previously in his

"Biografía de Tadeo Isidoro Cruz" ("Biography of Tadeo Isidoro Cruz") and especially "El fin" ("The End"), an alternative denouement. The text operates as the gaps through which we visualize an alternative space and where the concept of the "carrera" (race)—an idea based in and erected atop the pillars of "order and progress," economic development, and national founding—is particularly challenged. And by doing so, the very notion of modernity itself is destabilized. The text as utopia, but utopia as protest, rebellion, and reformulation, as project and resistance. The utopia in *Las aventuras* is a plausible scheme for improvement to the extent that it opens our senses not so much to a linear rationality as to a perceptual phenomenology that enables us to reconfigure new ways of being. If there is something that the Anthropocene exacerbated, aesthetically speaking, it is the propensity to formulate, create, conceptualize, and imagine the potentiality that resides in the conceivable, and exit the "carrera" (race), as Cabezón Cámara suggests, to withdraw from "la máquina de tristeza e insatisfacción que nos consume" (the machine of sadness and dissatisfaction that consumes us) and, through resistance, bet on the potentiality of the unthinkable and unimaginable.[57]

Faced with ongoing deterritorialization and the appropriation of other people's lands—that is, a physical, cultural, and emotional ousting either to extract or "fill" (or both) the landscape—the transformation of the desert increases, turning it, along with the living communities that inhabit it, into purely exploitable and therefore disposable matter. The issue is no longer productive bodies. Technology has taken over, efficiently and effectively replacing them. Nor is it the place's unproductiveness. The desert space itself and the living bodies residing in its land have become expendable organisms, fit only for profitable purposes. Their materiality reverberates in the archives of extinction, archives built by those who became, as David Viñas aptly termed them, "los dueños de la tierra" (the owners of the land).[58] Union is intimate, yes, and within its foundations lie the fossilized remnants of a fenced-in landscape whose imaginary is gradually expanded through a web of voices that seeps into the cracked surface of the desert.

NOTES

Epigraph. Sarmiento, *Facundo*. All translations are taken from Kathleen Ross's translation.

1.   The scholarship on the Sarmiento-Alberdi dispute is quite extensive. Among the historians who have examined the subject, see, for example, Halperín Donghi (1980; 1994; 2005); Botana (1984; 1996); Terán (2008); and Hora (2001). On the internal contradictions within these positions and their transformation over time, see in particular Halperín Donghi, *Proyecto y construcción de una nación*.

2. Halperín Donghi, *Una nación para el desierto argentino.*
3. Some of the region's most representative mammals include Pampas deer, puma, Geoffroy's cat, Pampas fox, plains vizcacha, Brazilian guinea pig, nutria, and opossum.
4. Henni, *Deserts,* 11.
5. Henni, *Deserts,* 18.
6. Mires, *El discurso de la naturaleza,* 81.
7. Argentina's constitution was adopted a year later, in 1853.
8. Halperín Donghi, *Proyecto y construcción,* xxxiii.
9. Halperín Donghi, *Proyecto y construcción,* xx.
10. Sarmiento, *Facundo o civilización,* 225–44; Sarmiento, *Facundo: Civilization,* 228–50.
11. Sarmiento, *Facundo o civilización,* 227; Sarmiento, *Facundo: Civilization,* 231.
12. Sarmiento, *Facundo o civilización,* 227; Sarmiento, *Facundo: Civilization,* 231.
13. Sarmiento, *Facundo o civilización,* 234; Sarmiento, *Facundo: Civilization,* 239 (emphasis in the original).
14. Sarmiento, *Facundo o civilización,* 111; Sarmiento, *Facundo: Civilization,* 124 (my emphasis).
15. Sarmiento, *Facundo o civilización,* 112; Sarmiento, *Facundo: Civilization,* 125.
16. Sarmiento, *Facundo o civilización,* 112; Sarmiento, *Facundo: Civilization,* 125.
17. Sarmiento, *Facundo o civilización,* 215; Sarmiento, *Facundo: Civilization,* 220.
18. An exhaustive list of these men appears at the end of the text.
19. Sarmiento, *Facundo o civilización,* 218; Sarmiento, *Facundo: Civilization,* 223.
20. Scobie, *Argentina,* 32–33.
21. Sarmiento, *Facundo o civilización,* 218; Sarmiento, *Facundo: Civilization,* 223.
22. Scobie, *Argentina: A City and a Nation,* 70–71.
23. Sarmiento, *Facundo o civilización,* 34; Sarmiento, *Facundo: Civilization,* 58.
24. Sarmiento, *Facundo o civilización,* 34; Sarmiento, *Facundo: Civilization,* 58.
25. Viñas, *De Sarmiento a Dios,* 13. This expression had already appeared in issue 57 of *Crisis* (February 1988) in an essay titled "Sarmiento: Un gran burgués, ni beato ni perverso," and reprinted on October 9, 2021 (for the anniversary of his death). https://revistacrisis.com.ar/notas/sarmiento-un-gran-burgues-ni-beato-ni-perverso.
26. Sarmiento, *Facundo o civilización,* 239; Sarmiento, *Facundo: Civilization,* 245.
27. Krader, *Formation of the State,* 21.
28. In *La vuelta,* by contrast, more characters appear, tangents abound, and the tone is moralistic.
29. Hernández, *Martín Fierro,* 223; Hernández, *Martín Fierro the Gaucho.* All translations of Hernández (unless otherwise indicated) are taken from Kate Kavanagh's revised translation, published on her website (www.sparrowthorn.com).
30. Hernández, *Martín Fierro,* 192 (my emphasis); translation by Andrea Rosenberg.
31. Hernández, *Martín Fierro,* 213; Hernández, *Martín Fierro the Gaucho,* 14.
32. Hernández, *Martín Fierro,* 204; Hernández, *Martín Fierro the Gaucho,* 8.
33. Ludmer, *El género gauchesco.*
34. Quoted in Gori, *La pampa sin gaucho,* 53; translation by Rosenberg.
35. Hernández, *Martín Fierro,* 113; Hernández, *Martín Fierro the Gaucho,* 1.
36. Borges, *Obras completas (1923–1974),* 194; translation by Rosenberg.
37. Hernández, *Martín Fierro,* 226; Hernández, *Martín Fierro the Gaucho,* 22.
38. Hernández, *Martín Fierro,* 226; Hernández, *Martín Fierro the Gaucho,* 23.
39. Hernández, *Martín Fierro,* 244; Hernández, *Martín Fierro the Gaucho,* 34.

40. Hernández, *Martín Fierro*, 244; Hernández, *Martín Fierro the Gaucho*, 34.
41. Hernández, *Martín Fierro*, 245; Hernández, *Martín Fierro the Gaucho*, 34.
42. I use this expression following Graciela Montaldo in *Ficciones culturales y fábulas de identidad en América Latina*, 16.
43. Hernández, *Martín Fierro*, 373 translation by Rosenberg; Hernández, *Martín Fierro*, 372; Hernández, *The Return of Martín Fierro*, 63.
44. Viñas, *Indios, ejército y frontera*, 12; translation by Rosenberg.
45. Echeverría, *La cautiva - El matadero*, 42; translation by Rosenberg.
46. Echeverría, *La cautiva - El matadero*, 48; translation by Rosenberg.
47. Henni, *Deserts*, 12.
48. Henni, *Deserts*, 12.
49. Alberdi, *Luz del día en América*, 36; translation by Rosenberg.
50. Alberdi, *Luz del día*, 136–37; translation by Rosenberg.
51. Pellow, "Toward a Critical Environmental Justice Studies," 223.
52. Quoted in Donghi, *Proyecto y construcción*, 90; translation by Rosenberg.
53. Cobas Carral, "Sarmiento/Alberdi"; translation by Rosenberg.
54. Cobas Carral suggests that this idea, which Alberdi first lays out in *Cartas quillotanas*, is the same explanation he uses twenty years later in "El Facundo y su biógrafo" to declare Sarmiento a failure both as governor of San Juan and as president of Argentina.
55. Cosgrove, *Social Formation and Symbolic Landscape*.
56. Cabezón Cámara, *Las aventuras de la China Iron*. The protagonist's surname, Iron, plays with the Spanish word *hierro* (iron), which is derived from the older term, *fierro*.
57. "Con deseo de vivir: sustrayéndonos de la máquina de tristeza e insatisfacción que nos consume mientras consumimos y moldeamos nuestras vidas en la línea de montaje de la 'carrera.' ¿Hacia dónde deberíamos correr qué carrera? ¿Cómo terminamos acá? ¿Qué nos hace sentir más vivos? ¿Un dispositivo nuevo o respirar en un bosque? ¿Un aplauso o trabajar en conjunto para crear algo distinto, algo que nos haga vibrar de ganas de estar vivos?" (With a desire to live: withdrawing from the machine of sadness and dissatisfaction that consumes us as we consume and molds our lives on the assembly line of the "race." In what direction should we run what race? How did we end up here? What makes us feel most alive? A new device, or breathing in a forest? A round of applause, or working together to create something different, something that makes us thrum with eagerness to be alive?). "Estéticas del Antropoceno: Tres preguntas con Gabriela Cabezón Cámara," on *Hablemos, escritoras* (Episode 342, May 30, 2022), https://www.hablemosescritoras.com/posts/856; translation by Rosenberg.
58. David Viñas, *Los dueños de la tierra*; translation by Rosenberg.

BIBLIOGRAPHY

Alberdi, Juan Bautista. *Luz del día en América*. Librería la Facultad, 1916.

Alberdi, Juan Bautista, and Oscar Terán. *Escritos de Juan Bautista Alberdi: El redactor de la ley*. Universidad Nacional de Quilmes, 1996.

Botana, Natalio R. *Domingo Faustino Sarmiento*. Fondo de Cultura Económica, 1996.

Botana, Natalio R. *La tradición republicana: Alberdi, Sarmiento y las ideas políticas de su tiempo*. Editorial Sudamericana, 1984.

Borges, Jorge Luis. *Obras completas (1923-1974)*. Emecé, 1974.

Cobas Carral, Andrea. "Sarmiento/Alberdi: Apuntes para una polémica posible (o de cómo construir los esquivos destinos de la patria)." V° Congreso Internacional Orbis Tertius de Teoría y Crítica Literaria, August 13-16, 2003, La Plata. *Polémicas literarias, críticas y culturales*. http://www.fuentesmemoria.fahce.unlp.edu.ar/trab_eventos/ev.8/ev.8.pdf.

Cabezón Cámara, Gabriela. *Las aventuras de la China Iron*. Literatura Random House, 2017.

Cosgrove, Denis E. *Social Formation and Symbolic Landscape*. University of Wisconsin Press, 1998.

Echeverría, Esteban. *La cautiva - El matadero*. Editor digital: MuadDib 09.05.14.

Gori, Gastón. *La pampa sin gaucho*. Editorial Universitaria de Buenos Aires, 1986.

Halperín Donghi, Tulio. *El revisionismo histórico argentino como visión decadentista de la historia nacional*. Siglo Veintiuno Editores Argentina, 2005.

Halperín Donghi, Tulio. *Proyecto y construcción de una nación: Argentina, 1846-1880*. Biblioteca Ayacucho, 1980.

Halperín Donghi, Tulio, Iván Jaksić, Gwen Kirkpatrick, and Francine Masiello, eds. *Sarmiento, Author of a Nation*. University of California Press, 1994.

Heffes, Gisela. "Estéticas del Antropoceno: Tres preguntas con Gabriela Cabezón Cámara." *Hablemos, escritoras*. Episode 342, broadcast May 30, 2022. https://www.hablemosescritoras.com/posts/856.

Henni, Samia. *Deserts Are Not Empty*. Columbia University Graduate School of Architecture Planning and Preservation, 2022.

Hernández, José. *Martín Fierro*. Biblioteca Ayacucho, 1987.

Hernández, José. *Martín Fierro the Gaucho*. Trans. Kate Kavanagh. Sparrowthorn.com, 2008. http://www.sparrowthorn.com/MartinFierro_PART_ONE.pdf.

Hernández, José. *Part Two. The Return of Martín Fierro*. Trans. Kate Kavanagh. Sparrowthorn.com, 2008. http://www.sparrowthorn.com/MartinFierro_PART_TWO.pdf.

Hora, Roy. *The Landowners of the Argentine Pampas: A Social and Political History, 1860-1945*. Oxford University Press, 2001.

Krader, Lawrence. *Formation of the State*. Prentice-Hall, 1968.

Ludmer, Josefina. *El género gauchesco: Un tratado sobre la patria*. Editorial Sudamericana, 1988.

Mires, Fernando. *El discurso de la naturaleza: Ecología y política en América Latina*. DEI, 1990.

Montaldo, Graciela. *Ficciones culturales y fábulas de identidad en América Latina*. Beatriz Viterbo, 2004.

Pellow, David N. "Toward a Critical Environmental Justice Studies." *Du Bois Review* 13, no. 2 (2016): 221-36.

Sarmiento, Domingo Faustino. *Facundo: Civilization and Barbarism*. Trans. Kathleen Ross. University of California Press, 2003.

Sarmiento, Domingo Faustino. *Facundo o civilización y barbarie*. Biblioteca Ayacucho, 1993.

Scobie, James R. *Argentina: A City and a Nation*. Oxford University Press, 1964.

Terán, Oscar. *Historia de las ideas en la Argentina: Diez lecciones iniciales, 1810–1980*. Siglo Veintiuno Editores, 2008.

Viñas, David. *De Sarmiento a Dios: Viajeros argentinos a USA*. Sudamericana, 1998.

Viñas, David. *Los dueños de la tierra*. Editorial Losada, 1959.

Viñas, David. *Indios, ejército y frontera*. Siglo Veintiuno Editores, 1982.

CHAPTER 4

Cuba's Ciénaga de Zapata

Despoiled Landscapes and Biodiversity Conservation in the Long Nineteenth Century

LIZABETH PARAVISINI-GEBERT

Located south of the province of Matanzas, the 6,000 km² Ciénaga de Zapata—or Zapata Swamp—is the Caribbean region's most important and best-preserved wetland. As a place of spectacular natural beauty, the traditional translation of "ciénaga" as "swamp" does it a disservice, since in addition to its large areas of proper swampland (wetlands dominated by woody vegetation such as trees and shrubs), the vast Ciénaga encompasses large areas of salt marshes, estuaries, floodplains, mangrove forests, and Caribbean coastal zones. Its ecosystem diversity and richness of flora and fauna are unmatched in the archipelago: more than 900 plant species; 175 bird species, including the endemic Zapata wren, Zapata rail, and Zapata sparrow, as well as the wondrous bee hummingbird, the world's smallest bird species. The endangered Cuban crocodile is among the thirty-one reptile species to be found there, together with more than 1000 invertebrate species. Its extraordinary biodiversity, augmented in winter by sixty-five species of birds during their annual migration from North America, led to its designation in 2000 as a UNESCO Biosphere Reserve and as a Ramsar Site in 2001.[1]

I want to focus here on the Ciénaga's history of environmental decay amid political turmoil in the long nineteenth century, a period during which the swamp became a threatened space "in imminent danger of being completely drained," many of its endemic species vulnerable to extirpation or extinction.[2] I trace here some of the rare textual glimpses offered by those venturing into its interior, a forbidding backwater, difficult to reach and inhospitably challenging to penetrate, from the mid-nineteenth-century to the early decades of the twentieth, as the fate of this most vital of natural Caribbean landscapes and of the unique biodiversity it supports hangs in the balance. These texts include a letter written by a young José Martí during the nine months he spent living with his father on the edge of the swamp; the narrative of the fast decline of the Cuban Red Macaw population in its last bastion of Hanábana, a decline that highlights the swamp's deterioration in the late 1800s; the memoirs of J. A. Cosculluela, an environmental engineer (1918), and Pura de Armas (2008), and their focus on the subsistence cultures of the Ciénaga, from which emerges a portrait of the typical *cenaguero* as a degraded peasant; and Luis Felipe Rodríguez's novel *La conjura de la Ciénaga* (1923), which portrays the zone as emblematic of the ills facing an independent Cuba in the early decades of the twentieth century. My analysis ends with a brief coda about the success of the Cuban Revolution's conservation efforts, which led to the revitalization of the Ciénaga, the protection of its endemic species, and its concomitant designation as a UNESCO World Heritage site in the year 2000. The coda underscores the swamp's recovery from the verge of extinction as a habitat and ecosystem— and through its recovery the preservation of the hundreds of specifies of fauna and flora that now thrive in the waterlands.

   In April of 1862, after a prolonged and difficult period of unemployment, José Martí's father, Mariano Martí, secured an appointment as the Circuit Captain for the region of Hanábana in the Ciénaga, about ninety miles southeast of Havana. Despite his wife Leonor's strong opposition, he insisted on taking young José with him, interrupting his son's schooling. The new post required frequent correspondence and reports to his superiors, and Mariano—insecure about his writing skills—wanted his son with him as a clerk. He was also interested in giving his son, who at nine years of age had just completed his elementary education, valuable work experience that would allow him to seek employment as a clerk once they returned to Havana. What the sojourn provided young Martí instead was a nine-month immersion in life at the edge of the Zapata swamp at a time when it was a crucial liminal space between an expanding sugar dependent economy and the expanse of "half-floating vegetation of grass and reeds, with clumps of willows . . . [where] many of the pools [were] so choked with aquatic vegetation that they show no open water at all."[3]

In 1862, Matanzas was a crucial zone for sugar cane production on the island, with a substantial slave population. Mariano's main area of operations, however, was on the edge of the plantation zone, in the deeply impoverished area on the northern edge of the Zapata Swamp. American ornithologist Thomas Barbour, in his memoir *A Naturalist in Cuba* (1945), would describe the San Francisco plantation near Zarabanda, in the vicinity of Hanábana, as the staging point for his explorations of the Ciénaga in the last two decades of the nineteenth century. The extensive wetlands, as described by Barbour and others, was a space of wondrous beauty, a marvelous place for a young boy unaccustomed to rural life, home to flocks of hundreds of pink flamingos or glossy ibises and the endemic and highly aggressive Cuban crocodile (*Crocodylus rhombifer*). Both magical in its beauty and eerie in its dense strangeness, the wetlands provided a natural barrier to the expansion of sugar plantations, and most of the inhabitants of Hanábana were primarily subsistence farmers, foragers, crocodile hunters, and charcoal makers, simple uneducated people who offered young Martí his earliest model for rural subsistence and his first glimpse at the travails and hardships of a sustainable engagement with a harsh surrounding environment in danger of being consumed by the sugar plantation. At the time of Martí's sojourn in Hanábana, plans for draining the wetlands were already under discussion, as the discourse of "modernity" and advances in technology allowed for projects aimed at integrating its rich and well-watered soil into the plantation system while at the same time reducing the problems caused by flooding of which Martí would write to his mother, changes which in turn would make control of the coastline and the reduction of the illegal slave trade easier to achieve.

As Circuit Captain, Mariano's principal duty was that of stopping the illegal importation of slaves into the province. The instructions given to Circuit Captains stipulated their responsibilities to uphold laws against vagrants and drifters, control madmen who could disturb the neighborhood, authorize and participate in public meetings, and enforce all laws specified in the Government and Police Ordinances. Article 14 of the instructions referred specifically to the pursuit, capture, and return of runaway slaves and maroons in the district.[4] Mariano had secured his post in Hanábana precisely because of the reputation for incorruptibility he had gained as neighborhood warden in Havana. His unbending approach to the enforcement of the law and unwillingness to make allowances for class and racial hierarchies had not made him popular then, but they were the very qualities needed for the enforcement of the government's official policies on the slave trade. His predecessor had been released from his duties for engaging in precisely the very slave trading he had been there to impede.

The Hanábana district, with its difficult terrain and easy coastal access through Playa Girón (the Bay of Pigs), was ideal for furtive landings. Boats loaded with slaves would enter the Bay and move clandestinely through the vast Ciénaga toward the mouth of one of the intricate network of navigable rivers and channels in the wetlands, from which it would be possible to land their cargo and slaves on dry land under the protection of dense water-logged woods and mangroves.[5] Its proximity to the expanding sugar plantations in the Matanzas and Cienfuegos districts had made the swamp the center of illegal importation of slaves into Cuba. In 1862, however, the local government, under pressure from Great Britain, had committed to upholding the ban on the trading of slaves. That year, the United States had ended its participation in the slave trade and would abolish slavery on January 1, 1863, leaving Cuba as the only territory in the region still engaged in the African slave trade. The area under Mariano's jurisdiction had a substantial population of slaves and freed blacks already, with almost 900 slaves and around 350 free or semi-free individuals.

Conflicting interests at the highest levels of the colonial hierarchy would eventually make Mariano's position an untenable one; his uncompromising pursuit of his duties would lead quickly to troubles with his superiors. Slave trading—despite official agreements with Great Britain to work toward the imposition of laws against the trade—had been a steady source of income for the Spanish Regent Queen María Cristina, as for Cuba's Captain General and most provincial governors. The replacement of Captain General Gutiérrez de la Concha by General Francisco Serrano y Domínguez, Duke of La Torre, who had vowed to uphold the ban on the slave trade, had opened an opportunity for Mariano, but this opportunity would be short-lived. The naming of a new Captain General for Cuba, Domingo Dulce y Garay, Marqués de Castell Florit, in December 1862 sealed Mariano's fate, bringing his son José's rural adventure to an end. Dulce y Garay, who had married a wealthy Cuban plantation heiress, had property in the district and was not interested in a zealous and uncompromising Circuit Captain interrupting the flow of much needed slaves into the area. The zone of

Jagüey Grande, a small hamlet on the eastern end of Mariano's district—on the northeastern boundary of the Ciénaga—had been recently opened to sugar cultivation through extensive deforestation and was in dire need of manpower. The opening of Jagüey Grande to sugar cultivation was but the latest indication that the steady growth of the sugar industry south of Matanzas required the draining of the swamp and cutting of its forests—as well as an increase in the supply of illegal slaves as a time when Mariano had been quite successful in stemming the flow of illegal slaves within his jurisdiction. Judging by the records of christenings of adult slaves—a clear indication of new arrivals—there had been 225 adults christened in the local church in 1861. The numbers go down precipitously to twenty-three after the arrival of Mariano as circuit captain, only to go steadily up in 1863 after his departure, when he was replaced by the more accommodating officer he himself had replaced.[6]

Against this complex political and economic backdrop, young José's months in Hanábana represented one of the most unique formative experiences of his pre-adolescence years. It marked the beginning of a political and environmental education unattainable in his urban upbringing. Growing up in Havana, especially in the neighborhoods near the fortress of Las Cabañas and the port, he had grown up surrounded by urban black workers—artisans, skilled laborers, and domestic workers—free as well as slaves. His sojourn in Hanábana brought him into close contact with the conditions under which the work of enslaved black people sustained the plantation, a contact that would leave a lasting impression on the young boy, and of the toll that the expanding plantation system took on the Ciénaga's ecology, measured primarily through deforestation, rapid erosion, and the diminution of rainfall.

Mariano and his son reached Hanábana late in the cane season of 1862, just in time for the cane harvest, when the sugar production process moved from the fields to the mill. The voyage by train from Havana to Colón (then known as Nueva Bermeja) on the Güines rail line gave the boy his first glimpse of the vast fields of cane that supported Cuba's growing prosperity. From Colón they proceeded to Hanábana on horseback, as no other method of transportation was available, through roads that would become impassable in the fast-approaching wet season. As they would learn as soon as the rains began, once the Sabanilla River broke its banks, the small population of the district—some 3,400 people scattered across two villages and a handful of hamlets—would be isolated from the neighboring towns. It was, all in all, not an auspicious neighborhood. Jagüey Grande, the smaller of the two settlements in Mariano's new jurisdiction, was "un caserío tan miserable que apenas merece ese nombre" (a village so miserable as not to be worthy of its name).[7] Ornithologists studying the Cuban red macaw a few years before Martí's arrival in Hanábana described the area where the Hanábana River flowed into the swamp "as bordered by a wide area of open country with scattered clumps of palms and hardwood *cayos* [or] hammocks. The border zone sloped gently toward the swamp and its extent varied as the rains caused the water levels to rise or fall through the entire Ciénaga."[8]

Of the town where Martí lived with his father, there's nothing left now but the trace of the foundations of the bakery and the church and a small plantation of locust trees that marked the location of the cemetery. In Hanábana, they settled into the small wood house near the schoolhouse and the town's old church (a new one had been built in Jagüey in 1858 after the old one had been damaged by two consecutive fires). The house served both as headquarters and residence for the Circuit Captain. Its schoolhouse offered only the most basic education and could not accommodate the needs of young Martí, who had already mastered reading and writing and had excellent penmanship. The sojourn did offer him, however, his first eye-opening immersion into Cuba's rich natural world. Surrounded by cane fields to the north and the natural abundance of the Zapata swamp to the south, home to *cenagueros* making a precarious living as subsistence farmers and coal-makers, José threw himself into country pursuits, learning to ride horses and raise and train fighting cocks. We owe to Thomas Barbour a description of the landscape surrounding Hanábana some twenty-five or thirty years after Martí's sojourn in the town:

> Late in the afternoon I used to rest under some trees which allowed uninterrupted vision far out over the open marsh, and there watched for the bands of ibises which fed regularly along the drier shores, where cattle also wandered to eat the succulent hyacinths stranded along the marge. No man nor beast ventured far out on the *tembladera*, for this skin of vegetation rippled and sagged, and to break through meant oblivion. The ibises came regularly from the southwestern horizon in wavering lines, perhaps three or four hundred in a flock.[9]

José Martí's oldest known letter dates from his sojourn in Hanábana.[10] In an affectionate note to his mother dated October 23, 1862—the only example remaining of what appears to have been a frequent correspondence—he speaks of the joys of living in the countryside, where he has learned to ride horses and has discovered the rural pastime of cockfighting. The brief letter offers a surprisingly detailed account of the environmental conditions of the edge of the Ciénaga. He tells his mother of the problems they have faced sending and receiving mail because of flooding of the Sabanilla River, which impeded traffic to Nueva Bermeja (the old name for the city of Colón), the nearest train depot—difficulties typical of the early rainy season at the edge of the swamp. The flooding had affected nearby farms and had threatened two local farmers, Don Jaime and Don Domingo, but the waters had begun receding by the time he had finished his letter. Flooding was a perennial problem in the area because of its proximity to the wetlands. Indeed, the town of Hanábana would disappear by 1879 as people eventually relocated upland, with better access to the sugar plantations that offered work and better communication with larger towns.

His father, he reports to his mother, has been recovering well from a serious fall he had suffered while out in the floods. He has been taking excellent care of his horse, he tells his mother, fattening him like a pig: "ahora lo estoy enseñando a caminar enfrenado para que marche bonito, todas las tardes lo monto y paseo en él, cada día cría más bríos" (I am now training him to walk with a bridle so he can be ridden nicely, I ride him every afternoon, and he gets more spirited every day).[11] His horse absorbs most of his time and energy. The fighting cock he had received as a present from Don Lucas de Sotolongo, a local planter and slaveowner. Don Lucas, who belonged to a prominent local family with deep roots in the region, had several children of Martí's age, and had given him a *gallo fino*, a gamecock especially bred for fighting. Cockfighting is a sport with a long history and an enthusiastic following in the Caribbean, and this present marked Martí's initiation into this rural pastime. In his letter he declares it to be "very pretty" and speaks of the bond it has created between father and son: "es muy bonito y papá lo cuida mucho, ahora papá anda buscando quien le corte la cresta y me lo arregle para pelearlo este año, y dice que es un gallo que vale más de dos onzas" (Papá is taking very good care of it and is looking out for someone who can cut its comb and wattle and train it for me so we can take it to the cockfights this year, and says that it is a cock that is worth its weight in gold).[12]

Although Mariano did not encourage friendships between his son and the slaves and freed blacks living around them, it was difficult to avoid contact and he made friends among the rural blacks with whom he lived so closely, to whom he would allude years later:

¡Como olvidar a Claudio Pozo . . . yendo al potrero, hablando con su
negra . . . y sobre todo a Tomás; Tomás que era para mí el señor Tomás,
el Señor Don Tomás, el Excelentísimo Don Tomás, Su Majestad Tomás,
que lo era todo para mí, era mi amigo. Era bueno y tenía espíritu nuevo y
artístico. Me deleitaba cantando y silbando. Travieso con todos los demás,
quieto a mi lado.

How could I forget Claudio Pozo . . . going to the paddocks, speaking to
his negress . . . and especially Tomás: Tomás who was for me Mr. Tomás,
Don Tomás, the most excellent Don Tomás, His Majesty Tomás, who was
everything to me, who was my friend. He was good and had an original
and artistic spirit. He delighted me with his singing and whistling. Mis-
chievous with everyone else, he was quiet when with me.[13]

As he accompanied his father on long rides throughout the district, im-
mersing himself in the strange liminal terrain between the plantation and
the swamp, he came face to face with the crowded barracks, dreadful conditions, and corporal punishments meted out to slaves on the plantations.
"¿Y los negros?"—he would write years later "¿Quién que ha visto azotar a
un amigo y no se considera para siempre un deudor? Yo lo vi, lo vi cuando
era niño, y todavía no se me ha apagado en las mejillas la vergüenza" (Who
has ever seen a friend physically whipped and does not consider himself for-
ever in that man's debt? I saw it, I saw it when I was a child, and I can still
feel the shame burning on my cheeks).[14] Moreover, his work as his father's
clerk—which entailed drafting and copying Mariano's frequent reports to
his superiors—drew him into the world of the slave trade. His father's main
task was that of curtailing the debarkation of slaves in his district, and Martí
will describe vividly such an event in his *Versos sencillos*, where he recreates
the eerie solemnity of the deep swamp.

> XXX
> El rayo surca, sangriento,
> El lóbrego nubarrón:
> Echa el barco, ciento a ciento,
> Los negros por el portón.
>
> El viento, fiero, quebraba
> Los almácigos copudos;

Andaba la hilera, andaba,
De los esclavos desnudos.

Thunder plows, bloodstained,
Through the dark storm cloud:
The ship spills out the blacks
Hundred by hundred through its gates.

The fierce wind shattered
The thick-topped mastic tress
And out in a single file
Marched the naked slaves.[15]

The poem also narrates the dramatic incident of the poet's encounter with the corpse of a slave hanging from a *ceiba*, the ubiquitous silk-cotton tree sacred to both Cuba's Indigenous and African populations. The incident is often incorporated into biographies of Martí as a historical event. There is, however, no evidence of its historicity, despite the earnest tone of the narrative and the dramatic character of the child-witness' pledge:

Un niño lo vio: tembló
De pasión por los que gimen:
¡Y, al pie del muerto, juró
lavar con su vida el crimen!

A child saw it: he trembled
Passionately for those who weep
And, at the dead man's feet he swore
To wash the crime with his life![16]

Martí's later writings about slavery trace the roots of his fiercely abolitionist stance to these months he lived in Hanábana, his only continuous experience of rural living in Cuba. The love of nature that underpins his poems about the island from which he would be exiled throughout most of his life, the appreciation of the character and strength of its peasantry, the sympathy for an unjustly degraded and abused race, the connection to the island's flora and fauna, stem from these most significant nine months—a true gestation period for a budding patriot with an understanding of the environmental conditions determining the lives of Cuba's subsistence rural population. In Hanábana, Martí began his ideological separation from his Spanish father's identity as a military man, from his anti-Creole ideas, and intense loyalty to his native Spain. This process was aided by the relationships Martí developed in Hanábana with local Creole sugar planters and *cenegueros* living precariously off the swamp, who provided an invaluable education in the complexities of agricultural production and the obstacles to reform represented by Spanish colonial authorities and laws, especially their exploitative tax codes. The Creole planters he met were of the type that six years later would lead the fight for slave emancipation and political independence known as the Ten Years' War (1868–1878).

Martí's reflections on this initiation into the realities of the Cuban countryside can be read as linking his approach to the understanding of the Cuban proto-nation to earlier assessments of Cuba's emerging Creolité. Here I am thinking particularly of his readings of José Martín Félix de Arrate y Acosta's *Llave del Nuevo Mundo* (*Key to the New World*, 1761), where early notions of Cuba's identity are linked to a celebration of Cuba's environmental richness and to an implicit project for the conservation of this diversity against the forces of the Spanish empire which, deep into the eighteenth century, seek to move toward a mono-crop system, following the successful example of the French and British Caribbean colonies. The focus of Arrate's text is to underscore the difference, diversity, and implied self-sufficiency and sustainability of the island as a unique environmental system. His text is both a description and a boast of an enviable abundance that is the foundation for a proto-national identification, of an expression of an incipient *cubanía* that will begin the separation between producers of sugar living in the deforested plains and those able to plant and profit—yet still conserve and live—on the edge of the abundance and protection of the forests, planting cacao, coffee, and tobacco. The latter are precisely the types of rural planters (subsistence and otherwise) Martí encountered during his experiences in the Ciénaga. Arrate, in his effort to emphasize this difference, establishes a clear distinction between these lands of natural abundance and the "tierras de labor" where slaves work in sugar fields. In his narrative, Arrate implies the existence of three distinct groups in Cuban society— all with a different relationship to the landscape: the Spaniards who have brought irreversible changes to the landscape in the form of new crops and animals; the *isleños* or islanders (the peasantry that represents both Spanish and Indigenous ethnicities) whom he sees as belonging to the landscape in an especially authentic way as they live off the bounty of the forests, wetlands, and sea; and those like him—upper-class and Cuban-born but not Indigenous—inextricably connected to a new definition of the landscape, seeking to profit from the landscape without altering it inexorably. Martí's environmental imagination will be rooted in Arrate's understanding of what was "natural" for Cuba and America.

It is difficult to glean the accuracy with which the very young Martí could "read" the Ciénaga's landscape and assess the nature of its harmonious sustainable rhythm—the "logic of relation" is the term he uses—as he does in "Livingstone," a travel article about his voyage through the coastal zone of Yucatán in which he describes the Indigenous inhabitants of a coastal village, whose lives and relationship to nature echo those of the *cenagueros*:

Estas caribes de opulento seno son las cultivadoras de los campos; los hombres pescan y comercian; las mujeres siembran y hacen su oficio de madres y de esposas. Las mismas manos que introducen en la tierra el *vástago*, le arrancan luego su raíz jugosa, y lo brindan luego al viajero en ancha torta. Son admirables esta vivacidad, esta generosidad, esta fraternidad, esta limpieza. . . . Es un pueblo moral, puro, trabajador.

These Carib women of ample breasts work as planters in the fields. The men fish and trade; the women tend the fields and fulfill their duties as wives and mothers. The same hands that plant the crop will pull it by its roots, squeeze its juices and serve it later to the traveler as bread. Their vivacity, generosity, and fraternity is admirable. . . . It is a moral, pure, hard-working town.[17]

In "Our America," Martí builds his arguments on a highly nuanced—and at times ambivalent—deployment of references and metaphors on nature and natural man. Central to these references is the notion of the land and its potential abundance as being "lo que la Naturaleza puso para todos en el pueblo que fecundan con su trabajo y defienden con sus vidas" (what nature made available to all nations to foster with their work and defend with their lives).[18] "Natural man's" knowledge and careful husbanding of this natural bounty is the key to the nation's prosperity, but man has to be careful to manage the land according to the needs of its people as the only means of avoiding tyranny. Good government, Martí writes, follows the

"soul of the earth" and avoids the mistakes of the past—"los elementos discordantes y hostiles que heredó de un colonizador despótico y avieso, y las
ideas y formas importadas que han venido retardando, por su falta de realidad local, el gobierno lógico" (the discordant and hostile elements inherited
from its perverse, despotic colonizer with the imported forms and ideas that
have, in their lack of local reality, delayed the advent of a logical form of
government).[19] We can't truly trace the complexity of Martí's thoughts on
the balance between development and sustainability for the Cuban proto-
nation to a brief sojourn at the Ciénaga de Zapata. But it is possible to see
that salient episode in Martí's life as the seed of an understanding of Cuba's
options that would sprout through observations of other Caribbean spaces
as having offered a glimpse at a spectrum of possible models of development in an agricultural Cuba. By design or necessity, it is a path Cuba has
followed throughout its complicated twentieth and twenty-first centuries'
history, culminating in the island's current efforts at approaching the kind
of sustainable development Martí envisioned in his writings on Mexico
and Guatemala—which found their echo in "Our America." Perhaps Fidel
Castro was right in claiming that Martí was "el autor intelectual del 26 de
julio" (the intellectual author of the 26th of July) and that the ideas behind
the Cuban Revolution find their most articulate expression in Martí's political thought.[20]

There is a poignant coincidence between Martí's sojourn in Hanábana
in 1862 and the retreat of the Cuban red macaw (*Ara tricolor*) from the area
into the interior of the Ciénaga, on its path to inevitable extinction within
the next two decades. The *Ara tricolor*, the only endemic local macaw, was
native to Cuba and the Isla de la Juventud (formerly Isla de Pinos, where
Martí spent two months under house arrest before leaving Cuba for Spain
as an exile in 1870).[21] Abundant at the moment of European arrival despite
being hunted and traded by the Indigenous population, the species began
its inexorable path to extinction immediately after 1492, when large numbers were captured for the European pet trade and its population began to
decline steadily as their forest habitats were destroyed to make way for the
expanding sugar plantation. By the 1850s, the macaws had retreated to the
northern range of the Ciénaga, where macaws "from the last bands" were
regularly seen feeding in a small group of paraiso trees "in the yard or *batey*
of the *colonia* at Zarabanda."[22] By the time Martí reached Hanábana, rapid
deforestation to clear ever more land for sugar cultivation had pushed the
macaw to the verge of extinction, with only a few breeding pairs remaining

in the environs of Hanábana, where we can tantalizingly imagine the patriot and the last remaining individuals of a species having lived in the Caribbean islands for as many as twenty-five million years briefly encountering each other. The species is believed to have survived in the most remote areas of the Ciénaga de Zapata until the late 1880s. The last pair seen in Isla de Pinos was shot in 1864 at San Francisco de la Vega, six years before Martí arrived on the island. The *Ara tricolor* was the last surviving species of at least seven (or as many as fifteen) macaws living in the archipelago at the time of the European encounter.[23] Thomas Barbour, during his stay in Hanábana, expressed his concern over the impact of the plantation into the Ciénaga not only through his concerns for the Cuban red macaw, but through his fear for the local extirpation of the glossy ibis (*Plegadis falcinellus*): "if ever cane is planted in the Ciénaga, the last chapter will be written in the Cuban history of the glossy ibis."[24]

The narrative of the macaws' imminent extinction, of which Martí was surely not aware, is nonetheless deeply linked to the political and economic moment of his seminal experience in the Cuban countryside. The very same conditions that led to the macaw's disappearance—the expansion of the plantation system to meet international demand, the concomitant deforestation, the tightening of Spanish economic control aided by the profits from sugar cultivation, the increased pressure on the rural population to join the plantation labor force, Spanish adherence to the slave trade at a moment when it has been abandoned by most European nations and banned throughout the archipelago—would all lead directly to the Ten Years War and Martí's own deep politization. The juxtaposition of the two narratives underscores the environmental cost of market-driven colonial exploitation, in this case through an extinction linked directly to the expansion of the plantation and on the pressures that expansion imposed on the surrounding wetlands. The local response—the Ten Years' War—was driven openly by perceptions of these costs by Creole farmers and the plantation workers (slaves, former slaves, indentured servants) who would join in a prolonged war of independence against Spain motivated, in part, by the desire to regain control of the environment and establish a more sustainable relationship with the land.

Independence, however, will open a problematic period for the Zapata Swamp, as the late nineteenth-century's focus on modernity, together with capitalism's emphasis on control and exploitation of natural resources, brings forth the question of how to solve a "problem" like that of the pre-modern

FIGURE 4.1. *Red Cuban Macaw* by Louis Agassiz Fuertes. Image courtesy of the Macaulay Library at The Cornell Lab of Ornithology.

Ciénaga. Accustomed to land being valued in terms of its productivity, the "unproductive" swamp becomes the focus of a variety of "modern" government-sponsored projects based on dubious scientific principles whose aim was to foster the economic viability of an area deemed economically, political, and socially backward. With no understanding at the time of the number of vital ecosystem services the wetlands provided, with the lowest population density and highest levels of illiteracy of any Cuban district, and with economic activity limited to subsistence agriculture and charcoal making, the plans focused primarily on the draining of the swamp through the development of a complex network of canals. These would allow—or so it was thought—for the expansion of the sugar industry into the drained lands, which, in addition to the production of hemp, the processing of bagasse for papermaking, cattle-raising, and fishing, would provide a solid foundation for a new economy in the region. Unvalued and unprotected were the region's unique biodiversity, ecosystem services, and importance as a waterfowl habitat, all environmental assets still not sufficiently valued by science or capable of being positively assessed by economists.

We are fortunate to have a vital chronicle of this period in *Cuatro años en la Ciénaga de Zapata* (1918), a richly textured report by a young engineer who spent four years in the Ciénaga in the early years of the century as part of a team of technicians in charge of evaluating the draining project and the development of new industries and communication infrastructure. The "report" on his failed engineering project, focused on the need for modernization in the swamp, offers nonetheless a colorful albeit deeply biased social, geological, and anthropological history of the swamp with a richly nuanced ecological assessment, even if his "modern" conclusions would point to the destructive transformation of a unique environment.

The report, nonetheless, as Reinaldo Fuentes Monzote argues in his study of deforestation in Cuba from 1492 to 1926, became a valuable document for understanding the environmental impacts of earlier economic development such as the expansion of the plantation into areas unsuited to sugar production, as witnessed by a young José Martí in the early 1860s. These incursions, Cosculluela chronicles, had a detrimental impact on the lands on the northern edge of the Ciénaga, where Hanábana was located: patterns of rainfall were changed and "flooding swept the lands, drowned the cattle, and caused an infinitude of evils," as Martí himself had experienced.[25] These very conditions would lead to the abandonment of the town some decades later. Of particular interest is Cosculluela's account of the tug-of-war between local sugar planters and the Spanish Navy for control of local forests. His understanding of hydrology allows for unique insights into the patterns of water flow and their impacts on local populations, which, together with his observations about the impact of deforestation on soil conditions and rainfall, attests to the pressures on the swamp's fragile environment. Moreover, the text is of deep interest in its bringing together, under a proto-environmentalist lens, current scientific knowledge from leading naturalists, anthropologists, archeologists, and sociologists. Legends and traditional tales abound, guided by a fascination with Taíno mythology and the history of piracy, and which offer an understanding of both human impact on the swamp and of the ways humans coped with the wetland conditions across centuries.

The report, moreover, opens a new narrative vein that will persist to this day, despite the deep social transformation undergone by the region after the Cuban Revolution of 1959: that of the *cenaguero* as *monteros,* a special group of Cubans deeply identified with the bush, inhabitants of the "Africa Tenebrosa de nuestra tierra, region misteriosa, temida, fantástica y poblada

por seres que en su completa orfandad, perdieron el contacto con la civilización y el progreso, manteniéndose estancados en su cultura, y en un estado
de barbarismo casi completo" (the darkest Africa of our land, a mysterious,
feared, and fantastic land populated by beings who, in their comprehen-
sive orphanhood, lost all contact with civilization and progress, remaining mired in their culture, and in an almost total state of barbarity).[26] It is
a narrative of particular relevance in our context, as it portrays the human
population as subsumed by the natural environment, as one and the same,
humans and wilderness as fundamentally entwined. The persistence of this
narrative can be seen in a more recent text, *Historias contadas por Pura*, an oral
testimony of events that transpired a mere decade after the publication of
Cosculluela's book, as recorded and edited by Ernesto Chávez Álvarez in
2006. Pura's engaging text gathers the experiences of a peripatetic childhood
moving across the small, isolated settlements of the Ciénaga, as the family's
fortunes allowed. The daughter of a charcoal maker whose meager earnings
were supplemented by a small *conuco* or kitchen garden, hunting, and fishing, her family's life transpires in deep engagement with the Ciénaga's rich
natural world, from constant flooding to the scourge of painful and dangerous chiggers. The simple narrative—created through a series of interviews
by Chávez Álvarez—does not rise to the nuanced narrative style created at
the height of the testimonial literature trend of the 1960s and '70s, which
gave us Miguel Barnet and Esteban Montejo's *Biografía de un cimarron* (1966)
and Elena Poniatowska *Hasta no verte Jesús mío* (1969).

It does, however, capture the vicissitudes of a style of life in the Ciénaga
that had changed little since Martí's brief sojourn in 1862: a precarious existence based on sugar cane work and the unsustainable exploitation of the
mangrove forests to make the charcoal used locally for cooking, and marked
by the absence of education, medical care, and the most basic hygiene. From
Pura's simple descriptions we glean the burdens of a sustainable existence
spent in "un bohío destartalado en medio de la Ciénaga, desprotegido de las
lluvias, el viento y el frío" (a dilapidated *bohío* in the midst of the swamp,
defenseless against the rains, the wind, and the cold) and dependent for
food on "lo que la naturaleza a nuestro alrededor podía brindarnos' (what
the nature around us could offer us)—chiefly turtles, Cuban gar, crocodile, crabs, hutias, and "todos los pájaros de la Ciénaga" (all the birds from
the Ciénaga).[27] Her subsistence existence is far removed from the ideals of
a bucolic life in communion with nature, plagued, as it was, by lice, chiggers, intestinal parasites, malaria, conjunctivitis, and hemorrhagic measles,

all treated without the aid of doctors through natural remedies found in the swamp, knowledge transmitted through generations to assure the survival of its population. The trajectory of her difficult childhood as the daughter of very poor parents for whom the Ciénaga was a last and generous refuge came to an abrupt end when her mother and six siblings were evicted by the forces of General Gerardo Machado after her father was jailed for being openly critical of his government.

Yet her recollections of the Ciénaga so many decades after having left are marked by a profound memory of the beauty and bounty of its vegetation, fauna, and generosity to those eking a living out of its richness. This beauty and bounty—as well as the travails and burdens—are expressed creatively by both Pura and her father through their writing of *décimas*, an often-improvised style of poetry common to rural communities in Latin America. From her father, "un poeta de tierra adentro que siempre tenía una décima flor de labios" (a poet of the land with a *décima* always on his lips), Pura learned the patterns and subject matters characteristic of the folk form, and her testimonial is peppered with her compositions and those of her father, a poetry born of their lives in the Ciénaga.[28] "Pues cenaguera nací," Pura writes in one of her *décimas*, "Y solamente yo vi / Fango, montes u lagunas, / Porque pañales ni cunas / Eso nunca conocí" (I was born a Cenaguera / and I only saw / Mud, mountains, and lagoons / since of diapers or cribs / I knew nothing).[29] Yet, the simplicity of her approach to the Ciénaga in her writing—a simplicity inherent in the form of the *décima* itself—matches her candid assessment of the beauty and richness of the natural world around it. Rich and generous, its harshness stems from the rigid rules of its ecosystem, one naturally indifferent to human life in its ecological imperative. It simply is, naturally, and humans must adapt to its rules and quirks.

It is an approach that differs greatly from that of Luis Felipe Rodríguez's *La conjura de la Ciénaga* (1926), a novel in which, as Gustavo Pérez Firmat argues, "the swamp is both setting and symbol . . . and realistic description and allegorical significance are glued together with a stickiness whose most accurate representation is perhaps the swamp itself."[30] Instead of Pura's portrait of "primitive and picturesque humanity" we get a Ciénaga whose ecological realities are burdened by the symbolic and heady turmoil of breathless *criollismo*. The plot is a familiar one: a jaded writer tired of the city's political cesspool accompanies his census-taking friend to the Ciénaga, where he plans to write a novel about the urban swamp. Confronted with the earthier and highly overheated passions of the real swamp, expressed primarily through an ardent sensuous woman and a young politician maddened by unrequited love and boundless desire, the Ciénaga destabilizes his moral and ethical compass. The young woman seduced, her would-be lover in quest for revenge, the city writer-turned-seducer is murdered, proving that civilization (in the guise of his census-taking friend) is powerless against the passions unleashed in the swamp where he finds his death:

> The monster was insatiable and perfidious. Nothing that entered its belly came out alive. It is said that it had swallowed me, animals, things. God take pity on someone who, unknowing, sets foot on that turbid and tranquil surface of still water! The swamp, obeying the spell of its infernal attraction, would devour him slowly, slowly, with irresistible and inevitable precision, until he disappeared from the light of the sun.[31]

There is more nuance to Rodríguez's novel than the description here may indicate, but it is not, in the present context, a novel that can be satisfactorily read ecocritically. Its Ciénaga belongs to the imagination as a generic swamp and does not further our understanding of the threatened space of Martí's and Pura's childhoods. In its engagement with the modernizing impulse that drives Cosculluela's own approach in *Cuatro años en la Ciénaga de Zapata*, however, the novel proposes a fitting ending to the long Cuban nineteenth-century's approach to the contested (and ultimately) remarginalized space of the Ciénaga. It will return to its dark and menacing obscurity as the pre-modern space that could swallow a talented and promising writer. Its ignorant and superstitious population, like the unfortunately seduced and disgraced Conchita, withdraws back into the swamp, "... como si la pasión invisible del malogrado Santiago Hermida la empujase desde el fondo profundo y oscuro de la ciénaga" (as if the invisible passion of the late Santiago Hermida pushed her from the bottom of the deep and dark Ciénaga).[32]

In March of 1959, not yet three months after the Revolution's triumph, Fidel Castro visited the "inhospitable" Ciénaga de Zapata for the first time, becoming acquainted with the coal makers' huts, the quality of the drinking water, the local population's experience with draining small fields for the planting of subsistence crops. A meeting with coal makers focused on a discussion of the canals built to reach the interior of the Ciénaga's "frightful solitude" and of the intricate processes of the artisanal production of coal and its long tradition in the area. Other meetings followed, including a memorable Christmas Eve celebration with the coal makers of the community of Sopillar in 1959. The celebration at the Ciénaga was part of a strategic plan to display his admiration for the "humble coal makers and peasants of the territory" (De Jesús) in the first holiday celebration under the new revolutionary government. The Ciénaga had been selected because it was deemed "one of the country's most backward territories" until the triumph of the Revolution —a vital part of the national narrative about the swamp. "Fue la primera vez en la historia de Cuba en que un dirigente de esa envergadura compartía con los trabajadores más explotados y olvidados . . . donde la gente vivía en paupérrimas condiciones, sin electricidad, sin médicos ni maestros y hacinados en rústicos bohíos" (It was the first time in Cuba's history that a leader of such significance spent time with the most exploited and forgotten workers, who lived in the poorest conditions, without electricity, without doctors or teachers, crammed in rustic *bohíos*), *Granma* exulted.[33] Some months later, in October 1960, Fidel and his companion Celia Sánchez visited the Ciénaga with French intellectuals Jean-Paul Sartre and Simone de Beauvoir, during what Sartre later called "the honeymoon of the Revolution."

These meetings and celebrations set into motion a number of initiatives for the improvement of lives and infrastructure in the Ciénaga of which Fidel would boast in a July 1961 speech to mark the inauguration of the Ciénaga de Zapata National Park:

> Esta era una zona completamente aislada del resto del país. . . . En dos años, esta zona se ha transformado totalmente. Hay actualmente en construcción tres pueblos para los campesinos de esta región; se han construido tres centros de recreación y descanso para trabajadores. Se está llevando a cabo un plan de repoblación forestal de vastas proporciones, y se ha construido cerca de 100 kilómetros de carreteras pavimentadas, y se ha construido más de 100 kilómetros de caminos. . . . Se ha organizado el Parque Nacional de la península de Zapata. . . . Pero lo más interesante es el trabajo que se ha realizado en esta zona en el aspecto humano. Actualmente, este 26 de Julio, el 80% de los analfabetos de la península de Zapata ha sido alfabetizado; es decir que la península de Zapata ocupa en este momento el primer lugar de Cuba, de las zonas rurales, en la campaña de alfabetización. Paralelamente, ocupa también el primer lugar de Cuba en la campaña de salud pública. Es el primer territorio de Cuba completamente vacunado.

FIGURE 4.2. Fidel at the Ciénaga de Zapata, December 24, 1959 (*Granma: Periódico Cubano*).

This was a completely isolated zone of the country. . . . In two years it has been totally transformed. Three villages are under construction for the people of the region, three recreation and rest centers for the workers have been built. A massive reforestation plan is underway and some hundred kilometers of highways and a hundred others of roadway have been built. A part of the swamp has been drained. . . . The Zapata Peninsula National Park has been established. . . . But the most interesting part of this work is what has been accomplished at the human level. Today, on this 26th of July, 80% of those who could not read or write in the Zapata Peninsula have learned to do so. That is to say that the Zapata Peninsula is now at the top of all rural regions in our anti-illiteracy campaign and occupies first place in the public health campaign. It is the first Cuban territory were all people have been vaccinated.[34]

By the time of the Park's inauguration, the Bay of Pigs (Playa Girón) invasion of April 1961 had brought intense national and international attention to the Ciénaga, adding to the wetlands' importance as a showcase for the profound changes brought about by the Revolution a new prominence as the site of the first foreign challenge faced by the revolutionary government. The failed invasion of April 17–20, 1961 was quickly repelled by Cuban forces, aided by the wetlands' difficult terrain and poor infra-structure. This new approach to an understanding of the importance of the Ciénaga as a national space required the reversal of the portrayal of the poor and illiterate swamp-dwelling *montero*. Turned into a protected national park with improved infrastructure and tourist facilities for local and foreign vis-itors alike, the Ciénaga emerged as one of the archipelago's greatest eco-tourist attractions, an international mecca for birdwatchers and nature lov-ers. Former charcoal makers were retrained as environmental officers, nature guides, or forest rangers; programs to protect endangered endemic species were developed using retrained local labor. Moreover, in 1993, this erstwhile secret, mysterious place was revealed to a broad Cuban audience through a limited series produced by Cubavisión. Focused on Ventura, an old *montero* from the Ciénaga, *Cuando el agua regresa a la tierra* (When water returns to the land) captures the moment when a small community of *cenegueros* is being moved to a newly built settlement less prone to flooding and closer to roads and modern infrastructure. Against the background of continued deforestation to produce charcoal and Ventura's wish to go deeper into the swamp to the spaces he had roamed in his youth, the series offers a contin-ued immersion into those hard-to-reach and -explore spaces that unveil the Ciénaga's hidden marvels. It is in part a travelogue through the wetlands and an introduction to its broad range of ecosystems. As an ecologically cen-tered narrative—perhaps the Caribbean region's first environmentally fo-cused melodrama—it is also a didactic attempt to teach the audience about the need for its preservation, bringing its focus onto the transfer of Ventu-ra's love and protection of the wetlands to a new generation through the couple of young lovers who stand for the Cuban audience. It remains one of the most beloved Cuban television series.

Capturing the enduring mystery of the Ciénaga, built on ineffable mo-ments of intense engagement between humans and a profuse and over-whelmingly enigmatic nature, continues to challenge Cuban writers and artists, as we see in an image from a documentary project by Cuban pho-tographer Raúl Cañibano, *Tierra Guajira*, begun in 1998 to create "ensayos generacionales que muestran la Cuba social" (generational essays that show the social Cuba) in which he grew up. Profoundly influenced by Brazilian documentary photographer Sebastião Salgado, Cañibano "va tejiendo his-torias alrededor de cada instantánea para finalmente, y como parte de un todo, reflejar modos de vida, costumbres y tradiciones que con el tiempo pueden o no perdurar" (weaves stories around each snapshot, and as part

FIGURE 4.3. Raúl Cañibano's *Man with Crocodile* (*Ciénaga de zapata*) from *Tierra Guajira* (2006). Photo courtesy of the artist.

of the whole, reflect on the ways of life, customs, and traditions that over time may or may not last).[35]

I cherish this photo for its nostalgic return to the nineteenth-century Ciénaga of crocodile hunters, the old struggles between *civilización y barbarie*, or for recalling Pura's affectionate descriptions of the pleasures of eating Cuban crocodile stew. It is an ambiguous, challenging photograph—a poacher? a member of the team working to protect the crocodile?—which reminds us of the ambiguities and challenges of environmental protection and conservation in sites such as the Ciénaga de Zapata. It reminds us that in the twenty-first century the safeguarding and protection of its ecosystems, fauna, and flora rests on the redirected skills of hunters turned conservationists, and that the Cuban crocodile—*Crocodylus rhombifer*, affectionately known as Castro's crocodile—is critically endangered and in dire need of protection.

Conserving the endangered Cuban crocodile was one of Castro's top priorities after first visiting the Ciénaga in 1959. His creation of a crocodile breeding facility in the Ciénaga was a central part of his commitment to the swamp and its endemic species. Yet the Cuban crocodile may be doomed, even if protected from hunters and poachers, by hybridization through mating with the American crocodile (*Crocodylus acutus*), present in Cuba and throughout the Caribbean archipelago and the surrounding mainland. Their fate will play out in the Ciénaga—now recovered as a UNESCO World Heritage site—and nature will take its course. The Cuban crocodile may be heading into extinction—as did the Cuban red macaws retreating from the sugar plantation and its deforestation when Martí spent those significant months living on the edge of the wetlands—but it is an exhilarating experience to come across one in the wild. I often wonder if the young José Martí may have been one of the last lucky ones to see the Cuban red macaw in flight against a clear luminous sky. As my colleague Alison Spodek Keimowitz wrote in a poignant essay about illness and climate change, loss may be inevitable in spaces as vulnerable as wetlands, "but the years we can postpone each loss, and each wild place and creature saved, are incalculably valuable."

NOTES

1. The Ramsar Convention, or the Convention on Wetlands, was established by UNESCO in 1971 as an intergovernmental environmental treaty to designate, conserve, and promote the sustainable use of wetland sites deemed to be of international importance, particularly those providing waterfowl habitats.
2. Barbour, *A Naturalist in Cuba*, 120.
3. Barbour, *A Naturalist in Cuba*, 120.
4. Florido Pérez and Fernández, "Martí en la Hanábana: Más allá de la aventura," 5–69.
5. Florido Pérez and Fernández, "Martí en la Hanábana," 27.
6. Florido Pérez and Fernández, "Martí en la Hanábana," 28.
7. Zéndegui, *Ámbito de Martí*, 29. See also Pezuela, *Diccionario Geográfico, Estadístico e Histórico de la Isla de Cuba*.
8. Barbour, *A Naturalist in Cuba*, 120.
9. Barbour, *A Naturalist in Cuba*, 120–21.
10. Hanábana, 23 de octubre de 1862. Estimada mamá: Deseo antes de todo que Ud. esté buena lo mismo que las niñas, Joaquina, Luisa, y mamá Joaquina. Papá recibió la carta de Vd. con fecha 21, pues el correo del sábado que era 18 no vino, y el martes fue cuando la recibió; el correo—según dice él—no pudo pasar por el río titulado "Sabanilla" que entorpece el paso para la "Nueva Bermeja" y lo mismo para aquí, papá no siente nada de la caída lo que tiene es una picazón que desde que se acuesta hasta que se levanta no le deja pegar los ojos, y ya hace tres noches que está así. Ya todo mi cuidado se pone en cuidar mucho mi caballo y engordarlo como un puerco cebón, ahora lo estoy enseñando a caminar enfrenado para que marche bonito, todas las tardes lo monto y paseo en él, cada día cría más bríos. Todavía tengo otra cosa en que entretenerme y pasar el tiempo, la cosa que le digo es un "Gallo fino" que me ha regalado Dn. Lucas de Sotolongo, es muy bonito y papá lo cuida mucho, ahora papá anda buscando quien le corte la cresta y me lo arregle para pelearlo este año, y dice que es un gallo que vale más de dos onzas. Tanto

el río que cruza por la "finca" de Dn. Jaime como el de la "Sabanilla" por el cual tiene que pasar el correo, estaban el sábado sumamente crecidos, llegó el de acá a la cerca de Dn. Domingo, pero ya ha bajado mucho. Y no teniéndole otra cosa que decirle déle expresiones a mamá Joaquina, Joaquina y Luisa y las niñas y a Pilar déle un besito y Vd. Recíbalas de su obediente hijo que le quiere con delirio. Martí, "Carta a Leonor Pérez y Cabrera," 43. (Dear Mamá: I hope, first of all, that you, along with the girls, Joaquina, Luisa, and Mamá Joaquina, are well. Papá got your letter on the 21st, as the mail didn't reach us on Saturday, the 18th, and it was Tuesday before it reached him; the mail—he says—could not be brought across the river called Sabanilla, to both Nueva Bermeja and us; Papá's not feeling any pain from the fall, but has been itching from the moment he goes to bed until he gets up, and has not slept a wink for the past three nights. All my attention goes into taking very good care of my horse and fattening him like a pig, I'm training him to walk with a bridle so he can be ridden nicely, I ride him every afternoon, and he grows more spirited by the day. And I have another thing to entertain myself and pass the time, and that is a gamecock that Don Lucas de Sotolongo gave me; it is very pretty and Papá's taking very good care of it and looking for someone to cut its comb and wattle and train it so we can take it to the cockfights this year, and he says that it is worth above two ounces. Both the river across Don Jaime's farm and the Sabanilla, which the mail has to cross, were very high, the one near us came almost up to Don Domingo's fence, but it has gone down a lot. I have nothing more to add other than sending my greetings to Mamá Joaquina, Joaquina, and Luisa and the girls and a kiss to Pilar and to you. Please accept it from your obedient son who loves you deliriously.)

11. José Martí, "Carta a Leonor Pérez y Cabrera," 43.
12. José Martí, "Carta a Leonor Pérez y Cabrera," 43. Gamecocks have their combs and wattles cut off in order to meet show standards and are conditioned so as to increase their strength and stamina. Martí refers to the gamecock as being worth at least 2 ounces, using the traditional weighing of gamecocks in Castilian pounds (worth 460 grams, divided in 16 ounces of 28.75 grams each).
13. Qtd in Zéndegui, *Ámbito de Martí*, 30.
14. Martí, "Carta a Leonor Pérez y Cabrera," 43.
15. Martí, "XXX," *Versos sencillos*, 42.
16. Martí, "XXX," *Versos sencillos*, 42.
17. Martí, "Livingstone," 38–39.
18. Martí, "Nuestra América." English version by Allen, "Our America."
19. Martí, "Nuestra América."; "Our America."
20. See Castro, *La historia me absolverá*.
21. Martí, sentenced to hard labor for sedition at the age of seventeen, was allowed to leave his prison in Havana to join a prison detail at "El Abra," a farm and marble quarry in Isla de Pinos, where he would remain for two months.
22. Ramsden, *Vida y exploraciones zoológicas*; and Barbour, *A Naturalist in Cuba*, 152.
23. Wiley and Kirwan, "The Extinct Macaws of the West Indies, with Special Reference to Cuban Macaw *Ara tricolor*," 125.
24. Barbour, *A Naturalist in Cuba*, 120.
25. Quoted by Fuentes Monzote, *De bosque a sabana*, 292.
26. Cosculluela, *Cuatro años en la Ciénaga de Zapata*, 327.
27. Chávez Álvarez, *Historias contadas por Pura*, 59, 67, 76.

28. Chávez Álvarez. *Historias contadas por Pura*, 37.

29. Chávez Álvarez, *Historias contadas por Pura*, 37.

30. Pérez Firmat, *The Cuban Condition*, 131.

31. Pérez Firmat, *The Cuban Condition*, 133.

32. Rodríguez, *La conjura de la Ciénaga*, 104.

33. García Gutierrez, "Fidel, en Nochebuena, con los pobres de la tierra."

34. Castro, "Discurso pronunciado por el Comandante Fidel Castro Ruz, Primer Ministro del Gobierno Revolucionario, en la clausura de los actos celebrados en Playa Girón, Península de Zapata, el 17 de julio de 1961."

35. Cañibano, *Tierra Guajira*.

BIBLIOGRAPHY

Barbour, Thomas. *A Naturalist in Cuba*. Little, Brown and Company, 1945.

Cañibano, Raúl. *Tierra Guajira*. Jíbaro Photos. https://web.archive.org/web/20201104181552/ https://www.jibarophotos.com/ongoing-projects/tierra-guajira.

Castro, Fidel. "Castro Speech at the Inauguration of the Ciénaga de Zapata National Park." *Revolución*. 28 July 1961.

Castro, Fidel. "Discurso pronunciado por el Comandante Fidel Castro Ruz, Primer Ministro del Gobierno Revolucionario, en la clausura de los actos celebrados en Playa Girón, Península de Zapata, el 17 de julio de 1961." Departamento de Versions Taquigráficas del Gobierno Revolucionario. http://www.cuba.cu/gobierno/discursos/1961/esp/f270761e.html.

Castro, Fidel. *La historia me absolverá* (1953). https://bnah.inah.gob.mx/bnah_lazaro_cardenas/uploads/E4_D124_FF1_18.pdf.

Chávez Álvarez, Ernesto. *Historias contadas por Pura*. Ediciones la Memoria, 2008.

Cosculluela, J. A. *Cuatro años en la Ciénaga de Zapata: Memorias de un ingeniero*. La Universal, 1918.

Florido Pérez, Rosario and José Fernández. "Martí en la Hanábana: Más allá de la aventura." *Anuario de Investigaciones Culturales*, vol. 7, 5–69. Grupo de Investigación y Desarrollo de la Dirección Cultural de Matanzas, 2007.

Fuentes Monzote, Reinaldo. *De bosque a sabana: Azúcar, deforestación y medio ambiente en Cuba, 1492–1926*. Siglo XXI, 2004.

García Gutierrez, Ventura de Jesús. "Fidel, en Nochebuena, con los pobres de la tierra." *Granma*, December 24, 2019. http://www.fidelcastro.cu/es/articulos/fidel-en-nochebuena-con-los-pobres-de-la-tierra.

Keimowitz, Alison Spodek. "I Felt Despair about Climate Change—Until a Brush with Death Changed My Mind." *Slate*, March 8, 2018. https://slate.com/technology/2018/03/an-environmental-professor-on-learning-to-cope-with-climate-change.html.

Martí, José. "Carta a Leonor Pérez y Cabrera," *Obras completas de Martí*, vol. 20. Centro de Estudios Martianos, 2001.

Martí, José. "Livingstone." *Obras completas de Martí*, vol. 19. Centro de Estudios Martianos, 2001.

Martí, José. "Nuestra América," *El Partido Liberal*, Jan. 20, 1891. http://www.josemarti.cu/publicacion/nuestra-america-2.

Martí, José. "Our America." *The Cuba Reader: History, Culture, Politics*, edited by Aviva Chomsky, Barry Carr, Alfredo Prieto, and Pamela Maria Smorkaloff, 119–24. Duke University Press, 2019. https://doi.org/10.1515/9781478004561-026.

Martí, José. *Versos sencillos*. La Veronica, 1939.

Pérez Firmat, Gustavo. *The Cuban Condition: Translation and Identity in Modern Cuban Literature*. Cambridge University Press, 1989.

Pezuela, Jacobo de la, *Diccionario Geográfico, Estadístico e Histórico de la Isla de Cuba*. Imprenta del Establecimiento de Mellado, 1863.

Ramsden, C. T. *Vida y exploraciones zoológicas del Dr. Juan Gundlach en Cuba (1839–1896)*. Sociedad Editora Cuba Contemporánea, 1918.

Rodríguez, Luis Felipe. *La conjura de la Ciénaga*. V. H. Sanz Calleja, 1924.

Wiley, James and Guy M. Kirwan. "The Extinct Macaws of the West Indies, with Special Reference to Cuban Macaw *Ara tricolor*." *Bulletin of the British Ornithologists' Club* (June 2013): 125–56.

Zéndegui, Guillermo de. *Ámbito de Martí*. Comisión Nacional Organizadora de los Actos y Ediciones del Centenario y del Monumento de Martí, 1953.

CHAPTER 5

Botanical Beings

On Women, Flowers, and Plants in Nineteenth-Century Latin America

VANESA MISERES

Women and Botany in the Nineteenth Century

From the end of the eighteenth and throughout the nineteenth century in particular, the link between women and botany acquired distinctive nuances in Western history. Ann B. Shteir affirms that the influence of Enlightenment science and the taxonomic urge to collect and systematize nature in all its variety gave botany cultural cachet and social sanction as an activity that combined amusement and self-improvement.[1] The simplicity of the Linnaean sexual system for naming and classifying plants according to the reproductive parts of flowers helped bring botany into prominence and gave it a little frisson at the same time.[2] Floral themes and designs became present in textiles, architecture, and porcelain, among other elements of daily life and culture. The proliferation of botanical art was driven mainly by women's enthusiasm for the subject. It was very common for elite and aristocratic women to approach the study of botany through the cultivation of plants and flowers, gardening, and drawing. Botany books and magazines written by and for ladies were widely published, especially in Britain.[3] Through botany, Victorian women discovered a language, a science, and an artistic outlet to express their creativity and intellectual abilities within the social restrictions that confined them to the domestic sphere.[4] Although many times this interest was primarily decorative — learning to draw in order to transfer those images to embroidery, for instance — in some cases women took this learning to the scientific field.[5] Others, as Jack Kramer explains, made a living illustrating flowers and gardens or writing about them, but they published mostly anonymously and did not achieve the recognition they deserved.[6]

The substantial corpus of texts demonstrating botany's influence on the literary works produced by Latin American women remains an understudied subject. Nevertheless, we know that botany occupied a significant space in the nineteenth century cultural, political, and economic spheres of the continent. Not only did the most relevant European botanists and geographers of the period visit and study the region (Alexander von Humboldt, Aimé Bonpland, or Alcide d'Orbigny, to name a few), but botany was also at the center of many debates on modernization and industrialization in the Latin American press. Botanical knowledge was mainly oriented toward agricultural purposes, and both intellectuals and scientists disseminated technical and scientific information for agricultural producers, who could use botany to modernize the economy of the continent.[7] Authors such as Domingo F. Sarmiento and Andrés Bello saw a path to progress in agriculture and botanical knowledge of their own regions.[8]

Women writers echoed this sentiment; Clorinda Matto de Turner wrote several journal articles on the importance of agriculture in the industrialization of her native Peru in the 1880s. In her novel *Sab* (1841), Gertrudis Gómez de Avellaneda includes comments on the sugar and coffee plantations to discuss slavery and the expansion of economic liberalism in Cuba. It also includes a lavish description of the garden Sab cultivates for Carlota. The garden becomes, on the one hand, a symbol of European taste and civilization (implied in the practice of horticulture) applied to the Cuban landscape. On the other hand, the presence of both native and foreign plants becomes a figure of transculturation and an expression of Sab's intelligence and aesthetic sensibility.[9] The Mexican periodical *Las violetas del Anáhuac* (1887–1889) is another clear example of the importance of botany among women. The four parts of "Algo de Botánica y Farmacología" (Something about Botany and Pharmacology) by Elvira Lozano Vargas, for instance, discuss a great variety of botanical themes considered useful for a female audience, from recommendations of plants and flowers according to the type of soil and weather available, to naturalist classifications of the vegetal kingdom, to the origin and properties of the main components of the Mexican diet, i.e., corn, rice, and barley.[10] Also in Mexico, José Joaquín Arriaga—founder of the Mexican Natural History Society—published "La vida de las flores" as part of his multi-volume project *La ciencia recreativa* (1871–1873). Although this publication was aimed at a non-specialist audience—specifically children, women, and the working classes—Arriaga dedicates this particular volume to his wife, Guadalupe Ponce de León de Arriaga, confirming the association between botany and female readers (Blanco no pagination). Puerto Rican feminist and suffragist Ana Roqué de Duprey (1853–1933) captured her interest in botany in her unpublished manuscript *Botánica antillana* (1933), in which she describes more than 6,000 species of plants and trees from her native island. This manuscript contains a mixture of poems and drawings, as well as aphorisms from the leading authors of the time. This combination provides a great source for a further analysis of the impact of botany on women's literary imagination and socialization.[11] Well into the twentieth century, Bertha Lutz (1894–1976) in Brazil and Helia Bravo Hollis (1901–2001) in Mexico personified the advancement of women as professional botanists in Latin America.

In this essay, I consider Latin American women's involvement with non-human nature through botany-related activities including the writing and drawing of flowers as safe and accepted forms of both socialization and the practice of science. While what was understood as "the language of flowers" was part of women's social and sentimental education, the practice of botanical drawing and writing also represented an opportunity to explore the scientific field that was usually reserved for men. I analyze the presence of botanical motifs in women's friendship albums as well as other women's literary works related to botany, such as María Josefa Acevedo's *El oráculo de las flores i el de las frutas* (1857) and Soledad Acosta de Samper's *Conversaciones y lecturas familiares* (1896). I show that the relationship between women and botany speaks of the barriers women faced in formal education and their attempts to overcome them in order to gain control of themselves and their own academic and social interests, as well as their sexual and reproductive bodies. Botany, although still unexplored as part of the Latin American "feminine world," can be recovered as an instrument for the construction of a social, sexual, and scientific history of women and their connections with their environments throughout the continent.

Friendship Albums and Botanic Drawings

Friendship albums represent a great example of how botany impacted Latin American women's social activities and pastimes during the nineteenth century. A typical album during this time consisted of a book bound with blank pages that served as a repository for collectible objects, images and writings, from musical notations, to drawings, dried flowers, poems, and postcards, among others. Located at the nexus of the public and private spheres, the albums were usually displayed in a parlor and even loaned out over the course of a few days in order to be inscribed by friends or renowned artists and poets of the time.

The album owned by Isadora Zegers, an interpreter and music composer of Spanish origin who emigrated to Chile, includes botanical motifs by both male and female authors from the 1840s until 1869. Botanical illustrations in her album do not represent female sensibility and emotion exclusively. "Through the medium of the senses," these illustrations connected art, science, and history and encouraged personal observation and self-education in the science of botany.[12] Pictures of plants were the medium for identification, analysis, and classification; they served not only the botanical scientist, the taxonomist, or the doctor, but also the plant collector, the gardener, the designer of applied arts, and the amateur enthusiasts of natural history.[13]

One of the drawings—in which I have not been able to identify the signature—illustrates the native Chilean bellflower, and it is accompanied by both the inscription of its scientific name, *Lapageria rosea*, and its native name *Copihue*. Strikingly, the story behind its European name is connected to women's interest in botany, since the name *Lapageria* is derived from the maiden's name (Lapagerie) of Napoleon's Empress Josephine, who was a keen collector of plants for her rose garden at Chateau de Malmaison. Copihue, on the other side, derives from the Mapundungun name (kopiw) which means "to be upside down," referring to the position of the plants' fruit, while the flower is called *kodkülla* in the Indigenous language.[14]

Copihue is revered as the national flower of Chile and is native to its coastal mountains and to the lower elevations of the Andes, in the southern third of the country where rainfall is relatively reliable throughout the year. The drawing in Zegers's album makes a reference to its geographical location with the annotation "Concepción." I like to think of this drawing, then, as providing the woman who owned the album with a window on both the local geography and flora and the global visual language of botany.

FIGURE 5.1. Illustration of *Lapageria rosea* in the album of Isidora Zegers. Courtesy of Biblioteca Nacional de Chile.

Sam George argues that botany was utilized in Britain to offer women a patriotic education, with the promotion of indigenous flowers over the exotic specimens that initially called the attention of British scientists. Such British attitudes to cultivated flowers were also dictated by theories of taste which, again, had ideological, nationalistic components.[15] Considering that Chile by the mid nineteenth century was a newly formed nation in the midst of political, cultural, and military struggles to impose a homogenous view of its national identity, we can imagine a similar pedagogical intention behind the drawing of a local flower that was given to a young woman who immigrated to the country at the age of twenty. Through the illustration of the *copihue*, the national, global, and historical contexts of science are connected

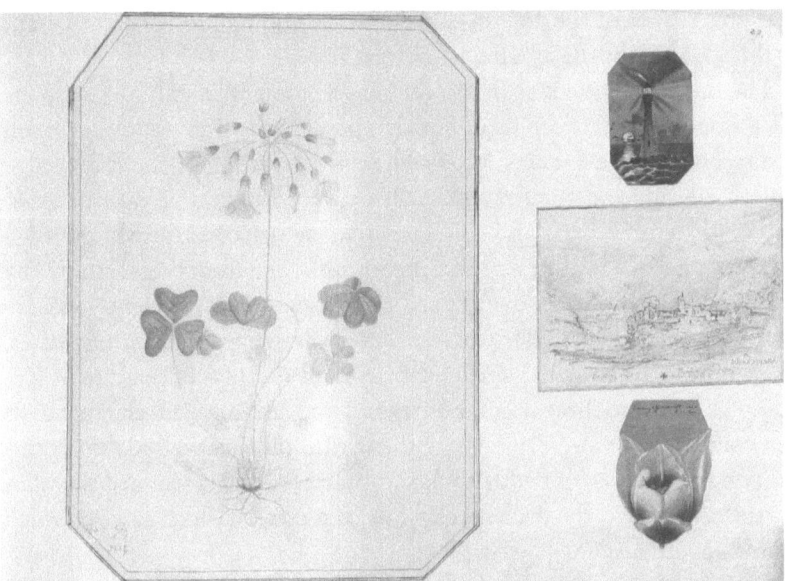

FIGURE 5.2. *Oxalis compressa* by Claude Gay. Album of Isidora Zegers. Courtesy of Biblioteca Nacional de Chile.

to the personal experience of her exile. Learning and becoming part of a new nation like Chile implies the exploration of its social networks and culture, but also its nature and its representation in art and science.

Another drawing from Zegers's friendship album is signed by Claude Gay in 1843 and it represents a wood-sorrel, or *oxalis compressa*, an ornamental flower native to South Africa. Claude or Claudio Gay (1800–1873) was a French botanist who carried out some of the first investigations of Chilean flora, fauna, geology and geography. Given his reputation as the "father" of Chilean botanical studies, it is curious to note that Gay gifted Isidora Zegers a drawing of a foreign flower. The album of a Chilean socialite gives him the freedom to illustrate something outside his typical duties. Offering Zegers an "exotic" flower might also be perceived as a pedagogical gesture intended to instruct women about other geographies and an opportunity for him to dialogue with the work of other European naturalists, such as the English Joseph Banks, who took part in Captain James Cook's first great voyage (1768–1771) and who first recorded *oxalis compressa* in 1760.[16] Although there are no specific botanical details in Gay's drawing, the habit and character of the plant are shown at the same time that the illustration keeps the balance of the design, expressing the willingness to create delight for the audience of the album.

The botanical drawings in the albums function as an exhibition of the cultural and economic capital of both the person who draws and the person who receives such drawings: they address the handling of the knowledge (that of painting and botany) that made up the education of women along with music, embroidery, and etiquette, all of which constitute knowledge of how to behave in different social contexts. The illustrations show the social status of the album owners and their friends, who evidently can acquire instruction manuals or pay teachers for drawing and painting classes, or have among their social circle established painters (as in the case of Zegers) who dedicate these floral motifs to them.[17] Albums and the materials they contained are also evidence of the leisure time, as well as discretionary spending, of the individuals who co-produced and consumed them. In short, they refer to the exchange circuit of a material economy in which friendships are sustained.[18]

The Language of Flowers: Women's Pastimes and Sexual Explorations

In nineteenth-century Europe and the Americas, botany was also in the center of the so-called "language of flowers" books. These volumes contained a list of flowers alphabetically arranged. Each flower was associated with a particular sentimental meaning such as hope, friendship, love, or mourning. As in the friendship albums, these books contain drawings with accurate representations of flower parts, stems, and roots. Some examples include the French author Charlotte de Latour's *Le Langage des Fleurs* (1819), a floral vocabulary that presents a symbolic system for the converse of lovers (Seaton 64); *The Flower Vase* (1849) by American author S. C. Edgarton, a collection of poetry and flowers according to popular meanings but with the inclusion of the botanical classification for each flower, as explicitly stated in the preface; and *The Coloured Language of Flowers* (1886) by English author Anna Christian Burke, which contains floral poems, a list of flowers and their assigned meanings, and illustrations.

In her *Oráculo de las flores y de las frutas* (Oracle of the flowers and fruits, 1857), Colombian writer María Josefa Acevedo de Gómez (1803–1861) rewrites the already codified language of flowers to adapt it to the women of Colombia.[19] The *Oracle* consists of two salon games of questions and

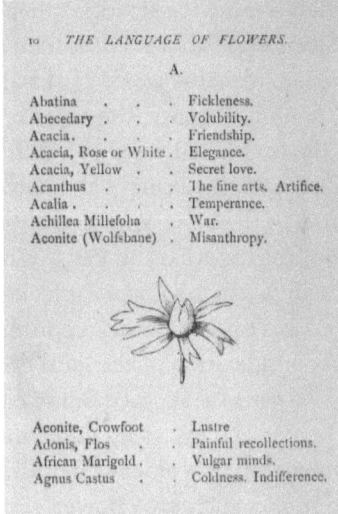
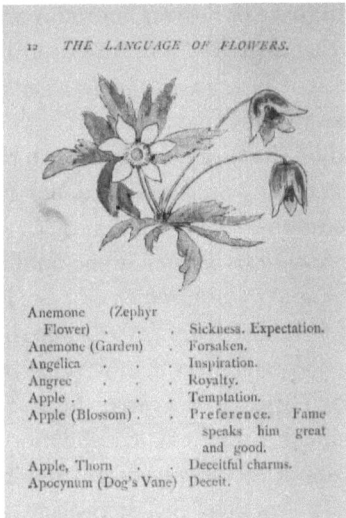

FIGURES 5.3 AND 5.4. Pages from *The Coloured Language of Flowers* by Mrs. L. Burke. George Routledge & Sons, 1886.

answers, one with flowers and the other with fruits. In the form of poetry, these "oracles" introduced questions about the future and provided answers that referred to literature, folklore, and superstition according to the numerical combinations resulting from dice thrown at random. As critic Ana María Agudelo Ochoa points out, María Josefa Acevedo must have been well aware, when she undertook the writing of her flower-themed book, that a genre so popular among women was likely to generate both revenue and social recognition as a professional writer.[20] That her desires were fulfilled we can deduce from the fact that the 1857 edition of the *Oráculo* was re-published as an addendum to another editorial success of the time, *El lenguaje de las flores i de las frutas*, published by Francisco de Torres Amaya.

There were numerous editions of *El lenguaje de las flores i de las frutas* in the Hispanic world, some of them dating back to 1846 and 1849 and some others appearing later such as a second 1857 version published in Bogotá, 1864 and 1882 editions published in New York, one in Barcelona in 1870 with several re-editions until 1913, and another one from 1889 that appeared in Mexico and Paris.[21] Each edition kept the same content structure with some changes and supplements that adapted the core of the book—an explanation of the language of flowers and how to use it—to their particular audience and region.

Language of flowers books involve, then, a transatlantic social phenomenon that includes Latin American and Spanish variations. For instance, all versions of *El lenguaje de las flores i de las frutas* explain the world as a balanced combination of masculine strength and feminine beauty, the latter represented by the flowers. But in each case, the text is addressed to a different audience: the *granadinas* or women from Nueva Granada in the 1857 Colombian editions, the South American women in the New York copy, or the Latin American women in the publication from Mexico and Paris. What is more, the 1857 Colombian edition, in which Acevedo's *Oráculo* is included, states that the flowers described in the book have been selected according to women's preferences in Colombia and in consideration of the local flora.[22]

The Hispanic editions of *El lenguaje de las flores i de las frutas* followed the traditional format of a compilation of poems and essays by different authors about the botanic world and its connection to women. They also included a "catalog" or "dictionary" section in which each flower was described and associated with a particular emotion—with explanations that refer to mythology, literature, and religion. Some of these entries included illustrations, and they all have a "Grammar" section in which the language of flowers is explained from a linguistic approach. Under this methodology, flowers and their shapes and positions are classified as nouns, adjectives, and verbs. The books established that women should learn these rules in order to subtly communicate their emotions through natural bouquets or floral drawings, which, at the same time, were expected to express women's good taste in their arrangement.[23]

Acevedo's *Oráculo* was intended to provide both edification and entertainment.[24] The questions proposed by the game not only refer to the language of affection that the flowers and fruits represented (love, friendship, and loyalty), but also included references to upper-class women's habits and concerns such as marriage, fitting into the social scene, or traveling to Europe. As Agudelo Ochoa affirms, this work aimed to regulate the behavior of readers through playful activities that appeal, in some cases, to fortune or chance.[25] In this sense the *Oráculo* played a role similar to that of etiquette manuals, which also influence the formation of citizens within a pre-established gender division.[26]

Acevedo's fruit and flower references are associated with desired values and unwanted behaviors of men, but even more of women (the intended audience for this book) in society. A large number of these representations such as fidelity, abnegation, modesty, and chastity are tied to

PREGUNTAS.	FRUTAS.	PJ.
1.ª ¿Debo creér en este oráculo?	*Frambuesa*	1.
2.ª ¿Qué fin tendrán estas cosas?	*Pasa*	2.
3.ª ¿En donde hallaré la felici-dad?	*Dátil*	3.
4.ª ¿Con quien me casaré?	*Fresa*	4.
5.ª ¿Qué debo hacér?	*Pomarrosa*	5.
6.ª ¿Cuántos años viviré?	*Nuez moscada*...	6.
7.ª ¿Cuál es mi pasion domi-nante?	*Naranjo dulce*...	7.
8.ª ¿Iré alguna vez a Europa?	*Uva de mar*....	8.
9.ª ¿Qué es lo que mas le gusta?	*Higo*	9.
10.ª ¿Variaré de pensamiento? ...	*Lima dulce*.....	10.
11.ª ¿Tendré razon de temer? ...	*Almendra*	11.
12.ª ¿Estaremos siempre en paz?	*Tomate*	12.
13.ª ¿Qué es lo que mas debo huir?	*Nuez*	13.
14.ª ¿Qué le contestaré?	*Pera*	14.
15.ª ¿Es bueno todo lo que agra-da?	*Mango*	15.
16.ª ¿Debo confiar mi secreto?	*Durazno*	16.
17.ª ¿Qué le diré que le agrade?	*Dominico*	17.
18.ª ¿En quien podré confiar? ...	*Guanábana*	18.
19.ª ¿Qué estado me convendrá?	*Toronja*	19.
20.ª ¿Qué haré si se muere?	*Mora*	20.
21.ª ¿Estará mi suerte aquí?	*Calabaza*	21.
22.ª ¿Hallaré lo que he perdido?	*Uva blanca*	22.
23.ª ¿Qué será de mí, dentro de un año?	*Granada*	23.
24.ª ¿Quién ahora mas me quie-re me querrá siempre?	*Ciruela*	24.

CHICAGO BOTANIC GARDEN, LENHARDT LIBRARY.

FIGURE 5.5. The list of questions and fruits included in Acevedo's *Oráculo de las frutas*. Imprenta de Francisco Torres Amaya, 1857.

women's sexuality. In the oracle, for instance, lilies represent innocence; white camellias, pure thoughts; the native flower known as *maravilla* (*Mirabilis jalapa* or the marvel of Peru), discretion; and dates, fidelity. At the same time, the verses that answer the players' questions advise women to avoid excess of any kind, to preserve morality, and to maintain a "gentle spirit" in marriage.[27] In the context of British literature, scholars have explored the connections between women's appeal to botany and women's sexual education. For Londa Schiebinger, the "scientization" of botany coincided with an ardent "sexualization" of plants; and the two phenomena cannot be understood in isolation. Linnaeus's plant taxonomy was based on traditional notions about (hetero)sexual hierarchies and in turn fixed normative gender relations through the scientific understanding of plants' sexuality. Although most flowers are hermaphroditic—that is, they have both female and male organs—Linnaeus and his successors understood plant reproduction through cultural metaphors based on the gender binarism of conventional marital relations; thus, anything "female" was conceived as a wife.[28]

In consonance with these developments around botany, Acevedo's *Oráculo* uses the language of flowers to teach women about sexual conduct. The idea of complementarity between men and women, the same that opened the *Lenguaje de las flores i de las frutas* and distinguished between female beauty and masculine strength, is stressed in the references to women as dependent on their ability to secure a marriage ("¿con quién me casaré?" [whom will I marry?]) or maintain a good reputation ("qué le diré que le agrade" [what would I say to please him/her?]).[29] According to Schiebinger, "complementarity provided an important ideological resolution to the new question of rights for women," at a time when traditional social controls were loosening, as can be seen in how marriage started to be celebrated not as a family arrangement but rather love between a couple.[30] Modern literature accompanied these changes with explicit descriptions of romantic and sexual encounters. In Colombia, these references to marriage and chastity through the language of flowers as a woman's duty and fate result from a particular interest in the context of the 1850s. In 1853, the liberal government ruling at the time passed a Civil Marriage Act (ley de matrimonio civil) that sought to oppose the atavisms of the colonial system, which even after thirty years of independence were still present in the daily life of the nation.[31] Although this law maintained the idea of women's dependence on men and their families, the conservative sector opposed this legal change in the concept of family. Mariano Ospina Rodríguez (1805–1885), founder of the Colombian Conservative Party and President of the country by the year the *Oráculo* was published, claimed that civil marriage represented a first step toward "the ideas of free love and of the community of women, that is to say, of the most unbridled libertinism erected in a system."[32] For Ospina Rodríguez, therefore, a society's future depended on the ability to control women's behavior through religious precepts of virtue and goodness. The Civil Marriage Act was repealed in 1856 and the ideal of a nuclear family continued to dominate social spheres, confirming the need for women to be "well educated" at marriage, in order to guarantee the existence of honorable families that would deliver healthy sons and daughters to the republic.[33]

In his poem "The Loves of the Plants" (1789), Erasmus Darwin, Charles Darwin's grandfather, uses botanical references to go beyond the limits of matrimony and reproduction in order to talk about a more plural concept of plants' sexuality. His verses present a botanic garden where pistils (female plants) seem to dominate the scene over stamen (male plants). Both groups of plants indulge in the pursuit of pleasure and relate to each other in several forms, not always following the "marriage" pattern of a pair.[34] Contrary to Darwin, who was a famously radical thinker, the language of flowers in Acevedo's oracle influenced and was influenced by strict moral and social expectations for women. Nevertheless, it also offered women the possibility to explore countless options within social, sentimental, and even sexual imagery. Several questions in the *Oráculo* express the interest of the (mostly) female audience in experiencing pleasure in forms that challenged the norms of 1850s Colombia. Players are offered the chance to ask, for instance, "¿es bueno todo lo que agrada?" (is all that is pleasing good?), "¿tendré pronto un gran placer?" (will I experience a great pleasure soon?), or "¿cuál de los tres es mejor?" (which of the three is better?).[35] Seeking answers in the language of flowers, women as well as men were able to verbalize more open concepts of love and sexuality within the bounds of upper-class sensibility and decorum. Playing with fortune while using both botany and the derived symbolic meanings of flowers, such questions and answers could be tolerated since they did not pose a serious threat to the social order.[36]

Soledad Acosta de Samper: Educating Women in Science

Another Latin American writer who reflected on the models of representation and sociability associated with botany was Soledad Acosta de Samper (1833–1913). Her father—the historian, scientist, and figure of Colombian independence, Joaquín Acosta—influenced her interests in history and science. The father and daughter duo shared a deep curiosity in a wide range of scientific and historical topics related to the natural history of Colombia. While serving in the Colombian army, Joaquín Acosta conducted a scientific and territorial survey of New Granada and in the 1840s, explored the western regions from Antioquia to Anserma, and wrote on Colombian topography, natural history, and native peoples. He also maintained a close relationship with the German naturalist Alexander von Humboldt, with whom he shared an interest in mining in the Chocó region. Soledad Acosta's botanical knowledge, on the other hand, is exhibited both as the expression of her female sensibility and as a manifestation of her professional activity as a writer, with the publication of several scientific pieces intended to circulate among female audiences and to improve women's education.

In *Conversaciones y lecturas familiares* (Family Readings and Conversations, 1896), the world of plants and flowers exceeds the sphere of feminine sensibility and is inserted directly into the field of science. As we have seen, botany was conceived as a discipline that fostered women's virtue and passivity. However, its study among women was limited to the purpose of creating a pastime and it was not recommended that women embark on large systems of classification, much less theorize about them—although there were women who did so.[37] On Latin American flora, we can cite—in addition to Soledad Acosta de Samper—the early case of Maria Sibylla Merian (1647–1717) and her analysis of the flora and insects of Surinam ("The Metamorphosis of the Insects of Suriname," 1705 and others), Maria Graham and her writings on Chile and Brazil (*Journal of a Residence in Chile During the Year 1822; and a Voyage from Chile to Brazil in 1823*, 1824), and Anne Kingsbury Wollstonecraft, with her manuscript on the plants and flowers of Cuba (*Specimens of the Plants and Fruits of the Island of Cuba*, 1823).

Conversaciones and lecturas familiares demonstrates Soledad Acosta's passion for and knowledge of botany and provides details and explanations of the medical and culinary uses of plants both native to Colombia and imported. The book is intended as an adaptation of the Victorian genre of *self-help* (training and educational books for young people), but in this case exclusively for women. Acosta fictionalizes a series of readings and lessons that take place on a farm on the outskirts of Bogotá. The owners entertain their visitors with lessons by a botanist and a priest who lecture on science and religion, respectively. Among the audience are two young women who, under the tutelage of the male figures, come into contact with scientific knowledge (here, again, Acosta seems to be well aware of gender restrictions). The classes occur during Sunday excursions on which the young women read passages from science books and spend most of the time discussing Linnaeus's system, biography, and theories. They cover the tea routes from Asia to Europe, the origin of spices such as pepper, vanilla, and cinnamon, and the importance of the Andean climate in the cultivation of potatoes and corn, among others. These conversations also highlight the contribution to botany of several English and North American women such as Mariana North and Febe Lankester, who serve as inspiration for the young Colombian women in and outside of the text.[38] Many of these passages had previously been published in the newspaper founded by Soledad Acosta herself, *El domingo de la familia cristiana* (Sunday of the Christian Family), under a section entitled "Nociones de botánica" (Botanical Concepts).

Acosta privileges here the connection between women and plants, and their uses from the scientific to the sentimental and the domestic. Her purpose in *El domingo de la familia cristiana* is to educate women as wives beyond the practical and everyday knowledge of things (plants in this case). It is a project, however, that has its limits, since it does not extend to all women. Lower-class women, for example, are represented in *Conversaciones y lecturas* as servants who prepare local dishes such as *ajiaco*, while the daughters of the landowners are the only ones who can venture into the world of botany.[39] It is also striking that Acosta's references to natural history are purely European-influenced, and thus the text gives no place to Indigenous knowledge or traditions about plants, among many others. At a time when South America is attempting to "civilize" itself along European lines, her approach to plants through science represents a desired connection to "progress." For this reason, Acosta's botanical discourse can be read as a gendered strategy to write and publish from a safe and accepted space that dares to show, nevertheless, the importance of women's scientific education in the construction of a modern national identity, a prospect that was not fully endorsed by the ruling classes.

To conclude, through the women's albums and other writings such as Josefa Acevedo's or Soledad Acosta's books, we can analyze the importance of botany as a feminine domain in nineteenth-century Latin America as well as the links among women, science, and the world of botany in this context. Those of us who work on the nineteenth century and, above all, women's literature, have generally focused on women's writings on education, about race and, to a lesser extent, about the world of work and the professionalization of women. However, the link between Latin American women and both science and plants, and the science of plants, has yet to be explored. Albums, as well as educational manuals, magazines, or women's columns are an excellent means to begin to examine the subject in greater depth. At a time when critics are debating the history of the Anthropocene, that is, the influence of human behavior on the earth, and revising the nineteenth-century canon to trace an eco-critical genealogy on the continent, it is important to incorporate the works analyzed here as gender inflections into the discussion.

The women I have studied and their involvement with plants speak, on the one hand, of their participation in a colonialist consciousness intent on controlling and possessing nature through Western science. But, at the same time, they allow us to decenter male voices in this context. Women, as less privileged subjects of nineteenth-century societies, offer alternative ways to engage but also criticize the narratives that have opposed plants to humans in a hierarchical discourse of domination and control. Their exploration of friendship, sexuality, and knowledge through a vegetal rhetoric and through concrete references to botany, let us think along a connection of women and plants beyond the anthropocentric and patriarchal point of view from which they depart. Following the ideas of Monica Gagliano, John C. Ryan, and Patrícia Vieira in *The Language of Plants*, we can recognize that these women's works listen to plants and speak to them in ways that are unquestionably mediated by culture and science, and that they do so in order to express *something else* about women and their environments. Plants and flowers expand the limits of gender relations, roles, and capacities. As Luce Irigaray puts it in *Through Vegetable Being*, women's communion with plants serves as a means of self-discovery and freedom "that can be gained only by oneself and with respect for others as living."[40]

Finally, if botanical drawings or literary, scientific, and cultural references to flowers and plants are a reflection of the souls and tastes of the young women of the nineteenth century, they do not capture an essence of the feminine but rather represent a search for other connections to the world that surrounds them, other ways of producing knowledge, not always governed by the predominance of the logos of writing, but also giving rise to other manifestations such as art and games. The relationship between women and botany reveals the constraints imposed on women's education and their attempts to evade or circumvent them in order to gain control over themselves and their environments, and to reveal the ideas that circulated in the nineteenth century about what is appropriate according to a binary distribution of gendered behavior.[41] If women were traditionally associated with the natural world of plants and flowers and, under the colonialist logic that invaded Western societies, they both became objects of control and domination, the differential relation that women propose with plants allows us to think of more powerful and empowering connections between them.[42] Among the work yet to be undertaken is the inclusion of non-Western women, whose experiences and knowledge can enrich and expand even further the study of nineteenth-century subjects' productive exchange with the vegetable world. Nevertheless, the examples provided with Isidora Zegers's inclusion of flower drawings in her album, Josefa Acevedo's reformulation of "the language of flowers," and Soledad Acosta's interventions in botany function as a vehicle to expand their intellectual and social horizon, as well as to revise the socio-sexual imagery for some women of nineteenth-century Latin America. Through botany women are rethinking their inclusion in humanity and, at the same time, as Irigaray expresses it, they are "elaborating a new education and sociocultural order," one that acknowledges, from nuanced perspectives, their bonds with other living beings.

NOTES

1. Shteir, *Cultivating Women*, 29.
2. Shteir, *Cultivating Women*, 29.
3. Kramer, *Women of Flowers*, 14, 41. Jack Kramer mentions, among other publications, Maria Jacson's *Botanical Dialogues* (1797) and Agnes Catlow's *Popular Field Botany* (1849), *Popular Garden Botany* (1855), and *Popular Greenhouse Botany* (1857).
4. Kramer, *Women of Flowers*, 13. Other studies have focused on British and American women's literature and botany, such as Elise L. Smith and Judith W. Page's *Women, Literature and the Domesticated Landscape*, Amy King's *Bloom: The Botanical Vernacular in the English Novel*, Deidre Lynch, "'Young Ladies Are Delicate Plants,'" and Kelli Towers Jasper, "Gathering Flowers: Romantic Era Botanico-Literary Production and the Transatlantic Mediation of Culture."
5. Shteir, *Cultivating Women*, 116.
6. Kramer, *Women of Flowers*, 14, 46.
7. Vega y Ortega Baez, "Botánica y agricultura," 170–71.
8. In "Plan combinado de educación común, silvicultura e industria pastoril" (Combined plan for common education, forestry and pastoral industry) from 1855, for instance, Sarmiento addressed his concerns related to the flora and fauna from Buenos Aires: the wild state of the herds, the lack of fences and watering systems, and the scarce development of agriculture. Likewise, Andrés Bello, in his poem "Silva a la agricultura de la zona tórrida" (1826), distinguished the physical elements of nature in order to establish the values on which a society should be based, in particular that of his homeland, Venezuela, and the whole American continent.
9. Wylie, *The Poetics of Plants*, 60–61.
10. For more details on women and botany in Mexico, see Torres Galán et. al., "En busca de la ciencia médica: De herbolarias a farmacéuticas en la ciudad de México, siglos XIX y XX," 73–97.
11. Long thought to have been lost, the manuscript was recently discovered by a group of researchers including Eliván Martínez and Jorge Carlos Trejo, and it is in the process of being digitized.

12. Saunders, *Picturing Plants*, 7.

13. Saunders, *Picturing Plants*, 7.

14. *Diccionario Mapuche*, 41, 155.

15. George, *Botany, Sexuality*, 159.

16. Druce, *The Flora of Oxfordshire*, 71.

17. Siegel, *Playing with Pictures*, 16.

18. Pelling, "Crafting Friendship."

19. I want to thank Nicolás Sánchez-Rodríguez for the reference of Josefa Acevedo's work during my presentation of a preliminary version of this essay at the MLA Congress held in Seattle in 2020. The first edition available of this book is 1857, but as a second edition; the first edition was from 1855, but is not available at this time.

20. Agudelo Ochoa, "Publicación de libros escritos," 165–66.

21. Agudelo Ochoa, "Publicación de libros escritos por mujeres en el siglo XIX en Colombia," 165; Pérez López, "Las mujeres y el lenguaje de las flores en la Barcelona de los siglos xix y xx," 314.

22. Acevedo, Oráculo, 10–12. Just to mention other examples of variation in the editions, the 1870 edition of the *Lenguaje de las flores* from Barcelona is signed by the pseudonym of Florencio Jazmín, name that alludes to the floral theme of the book. It also includes a text by renowned Spanish writer and journalist María de la Concepción Gimeno titled "Las niñas y las flores" (The girls and the flowers). In this text, the author refers to the language of flowers as a result of Linnaeus's ability to analyze the "psychology" of the plants and to "discover their loves" (18). The Mexico and Paris versions contain colored images of women's bodies disguised in different shapes of flowers, and the Paris edition added a section, "Lenguaje simbólico en la mesa" (Symbolic language at the table), which seems an abbreviated version of an etiquette book, a genre also popular among upper-class populations from Europe and the Americas (135–59).

23. Flowers with stem and leaves are nouns, an adjective will be the flower only (without leaves), and the verbs will be divided in present (open flower), past (flower with seed or without some petals), and future (flower with buds). The number of leaves, on the other side, will indicate the personal pronouns I / me (one leaf), you (two leaves), he / she (three leaves), we (four leaves) and they (five leaves).

24. In the sixteenth century, it was common to educate royal women and children through games of luck. See, for instance, Pedro de Guevara and his *Nueva y sutil invención en seys instrumentos, intitulado juego y exercicio de letras de las serenísimas Infantas Doña Ysabel y Doña Catalina de Austria, con la qual facilísimamente y en un muy breve tiempo se aprende todo el artificio y estilo de las gramáticas que hasta ahora se han compuesto y se compusieren de aquí en adelante.*

25. Agudelo Ochoa, "Publicación de libros escritos," 165.

26. Agudelo Ochoa, "Publicación de libros escritos," 165–66.

27. Acevedo de Gómez, *Oráculo de las flores*, 217, 221.

28. Schiebinger, *Plants and Empire*, 23–25.

29. Acevedo de Gómez, *Oráculo de las flores*, 213.

30. Schiebinger, *Plants and Empire*, 39.

31. Aristizábal, "La efímera existencia."

32. My translation, quoted by Aristizábal.

33. Aristizábal, "La efímera existencia."

34. Browne, "Botany for Gentlemen," 604.
35. Acevedo de Gómez, *Oráculo de las flores*, 213, 161.
36. Schiebinger, *Plants and Empire*, 33.
37. Shteir, *Cultivating Women*, 118.
38. Briggs, *The Moral Electricity of Print*, 140.
39. *Ajiaco* is a popular hearty soup from the Andean region of Colombia. It's made from three different kinds of potatoes (criolla, sabanera, and pastusa), chicken, guasca leaves, and corn.
40. Irigaray, *Through Vegetable Being*, 41.
41. Shteir, *Cultivating Women*, 114.
42. Plumwood, "Gender, Eco-Feminism," 44.

BIBLIOGRAPHY

Acevedo de Gómez, María Josefa. "Oráculo de las flores y de las frutas." In *El lenguaje de las flores y de las frutas con el oráculo por una señora granadina*. Imprenta de Francisco Torres Amaya, 1857.

Acosta de Samper, Soledad. *Conversaciones y lecturas familiares: Sobre historia, biografía, crítica, literatura, ciencias y conocimientos útiles*. Garnier Hermanos, 1896.

Acosta de Samper, Soledad. *El domingo de la familia cristiana*, 1889. Biblioteca Digital Soledad Acosta de Samper. https://soledadacosta.uniandes.edu.co

Agudelo Ochoa, Ana María. "Publicación de libros escritos por mujeres en el siglo XIX en Colombia: El caso de Josefa Acevedo de Gómez." *Cuadernos del CILHA* 18 no. 26 (2017): 155–76.

Aristizábal, Magnolia. "La efímera existencia del matrimonio civil en el siglo XIX." Red Cultural del Banco de la República, Credencial Historia, no. 269 (May 2012). https://www.banrepcultural.org/biblioteca-virtual/credencial-historia/numero-269/la-efimera-existencia-del-matrimonio-civil-en-el-siglo-xix.

Arriaga, José J. *La ciencia recreativa: Geografia descriptiva*. Aguilar, 1879.

Briggs, Ronald. *The Moral Electricity of Print: Transatlantic Education and the Lima Women's Circuit, 1876–1910*. Vanderbilt University Press, 2017.

Browne, Janet. "Botany for Gentlemen: Erasmus Darwin and 'The Loves of the Plants.'" *Isis* 80 no. 4 (1989): 593–621.

Burke, L. *The Coloured Language of Flowers*. George Routledge & Sons, 1886.

Diccionario Mapuche: Mapudungun/Español, Español/Mapudungun, 2nd ed., edited by Rafael Muñoz Urrutia. Editorial Centro Gráfico, 2006.

Catlow, Agnes. *Popular Field Botany*. Reeve, Benham & Reeve, 1849.

Catlow, Agnes. *Popular Garden Botany*. Reeve, Benham & Reeve, 1855.

Catlow, Agnes. *Popular Greenhouse Botany*. Reeve, Benham & Reeve, 1857.

Druce, George C. *The Flora of Oxfordshire: Being a Topographical and Historical Account of the Flowering Plants and Ferns Found in the County with Sketches of the Oxfordshire Botany During the Last Three Centuries*. Parker and Co., 1886.

Gagliano, Monica, John C. Ryan, and Patrícia Vieira, eds. *The Language of Plants: Science,*

Philosophy, Literature. University of Minnesota Press, 2017.

George, Sam. *Botany, Sexuality and Women's Writing 1760–1830: From Modest Shoot to Forward Plant*. Manchester University Press, 2017.

Guevara, Pedro de. *Nueva y sutil invención en seys instrumentos, intitulado juego y exercicio de letras de las serenísimas Infantas Doña Ysabel y Doña Catalina de Austria, con la qual facilísimamente y en un muy breve tiempo se aprende todo el artificio y estilo de las gramáticas que hasta ahora se han compuesto y se compusieren de aquí en adelante*. Herederos de Alonso Gómez, 1581.

Irigaray, Luce, and Michael Marder. *Through Vegetal Being: Two Philosophical Perspectives*. Columbia University Press, 2016.

Jacson, Maria. *Botanical Dialogues*. Joseph Johnson, 1797.

Jasper, Kelli Towers. "Gathering Flowers: Romantic Era Botanico-Literary Production and the Transatlantic Mediation of Culture." PhD diss., University of Colorado, Boulder, 2016.

Jazmín, Florencio. *El lenguaje de las flores y el de las frutas con algunos emblemas de las piedras y los colores*. Barcelona: Manuel Saurí, 1870; Paris: Bouret, 1889; Mexico City: Tipografía de Aguilar e Hijos, 1891.

Kramer, Jack. *Women of Flowers: A Tribute to Victorian Women Illustrators*. Stewart, Tabori & Chang, 1996.

King, Amy M. *Bloom: The Botanical Vernacular in the English Novel*. Oxford University Press, 2003.

Las violetas del Anáhuac (1887–1889). Hemeroteca Nacional Digital de México. http://www.hndm.unam.mx/consulta/publicacion/visualizar/558075bf7d1e63c9fea1a484.

Larsen, Claudia. "Botanical Art Treasures Stand Test of Time." Florida Wildflower Foundation, accessed July 11, 2021. https://flawildflowers.org/drawn-to-nature.

Lynch, Deidre. "'Young Ladies Are Delicate Plants': Jane Austen and Greenhouse Romanticism." *ELH: English Literary History* 77, no. 3 (2010): 689–729.

Miseres, Vanesa. "Lectoras, autoras y consumidoras: Los usos femeninos del álbum en Latinoamérica." *Revista Telar* 23 (2019): 25–48.

Pelling Madeleine. "Crafting Friendship: Mary Delany's Album and Queen Charlotte's Pocketbook." *Journal18*, no. 6 (October 2018), https://www.journal18.org/2909.

Pérez López, Fátima. "Las mujeres y el lenguaje de las flores en la Barcelona de los siglos xix y xx." *Revista Temas de Mujeres* 10, no. 10 (2014), http://ojs.filo.unt.edu.ar/index.php/temasdemujeres/article/view/85/84.

Plumwood, Val. "Gender, Eco-Feminism and the Environment." In *Controversies in Environmental Sociology*, edited by Robert White, 43–60. Cambridge University Press, 2004.

Saunders, Gill. *Picturing Plants: An Analytical History of Botanical Illustration*. University of California Press, 1995.

Schiebinger, Londa L. *Nature's Body: Gender in the Making of Modern Science*. Rutgers University Press, 2013.

Seaton, Beverly. *The Language of Flowers: A History*. University of Virginia Press, 1995.

Schiebinger, Londa. *Plants and Empire: Colonial Bioprospecting in the Atlantic World*. Harvard University Press, 2007.

Shteir, Ann. *Cultivating Women, Cultivating Science: Flora's Daughters and Botany in England, 1760–1860*. Johns Hopkins University Press, 1996.

Siegel, Elizabeth, ed. *Playing with Pictures. The Art of Victorian Photocollage*. Art Institute of Chicago, 2010.

Smith, Elise L. and Judith W. Page. *Women, Literature and the Domesticated Landscape*. Cambridge University Press, 2011.

Torres Galán, Josefina, et. al. "En busca de la ciencia médica: De herbolarias a farmacéuticas en la ciudad de México, siglos XIX y XX," *Letras históricas* 15 (Fall 2016–Winter 2017): 73–97.

Vega y Ortega Baez, Rodrigo Antonio. "Botánica y agricultura en la prensa argentina, cubana, colombiana y mexicana, 1822–1880." *Iberoamericana* 18 no. 69 (2018): 151–74.

Wylie, Leslie. *The Poetics of Plants in Spanish American Literature*. University of Pittsburgh Press, 2020.

Zegers, Isidora. *Album de Isidora Zegers de Huneeus*. Ediciones de la Dirección de Bibliotecas, Archivos y Museos, 2013.

CHAPTER 6

Hydraulic Energy, Nature, and Modernization in Nineteenth-Century Mexico

JORGE QUINTANA NAVARRETE

In February 2021, Mexico's president Andrés Manuel López Obrador launched a reform bill—later approved by the Federal Senate—that would foster state control in the energy market by prioritizing state-owned over private companies. This meant that the hydroelectric and fossil fuel plants owned by the state will have the opportunity to dispatch energy before private companies are allowed to. According to its proponents, this reform bill will still comply with the Paris Agreement that required Mexico to decrease its emissions and transition to renewable sources of energy. In this vein, President López Obrador unveiled a massive project to modernize and optimize more than sixty state-owned hydroelectrical plants in order to increase their production of clean and renewable energy. However, opponents argue that the reform bill will produce not only more expensive and less reliable energy by disincentivizing free and fair competition, but it will also deliver dirtier energy by undermining private-owned wind and solar energy companies, which are spearheading the transition to renewable energies. Since hydroelectrical plants are ageing and insufficient, environmental activists contend that the reform bill will ultimately bolster the state's reliance on fossil fuels, putting the country on a path inconsistent with the global fight against climate change and its socioecological consequences.

While there is still some reluctance to acknowledge the far-reaching effects of climate change, Mexicans are beginning to accept that the ever-increasing use of fossil fuels is not sustainable in the short term. At the same time, as illustrated by the debates around López Obrador's bill to reform the energy sector, it is by no means clear what comes after society's complete reliance on fossil fuels, and what sorts of political, socioeconomic, and cultural changes are expected to take place in the coming transition. The emerging field of the energy humanities has accepted the challenge of shedding light on the current impasse by exploring how the thorough dependence on fossil fuels came to exist in the first place and what the implications of its widespread and deeply ingrained influence on all spheres of society—from economic and political regimes to social relations of labor, cultural representations, subjectivities, structures of feeling, and everyday activities—are.[1] Terms proposed by energy historians and humanists—such as "carbon democracy" (Mitchell), "fossil capitalism" (Malm), "petromodernity" (LeMenager), "petrofiction" (Ghosh), among others—intend to critically examine the mutual interpenetration between modern industrial societies and fossil fuels' systems of extraction, circulation and use since the latter half of the nineteenth century. Energy scholars have afforded much critical attention to the historical period when the current fossil fuels-based structures were first put into place in the world, inasmuch as delving into past energy transitions—analyzing their negative socioecological impacts, as well as the new possibilities and potentialities opened by them—can provide us with conceptual tools to better understand and approach the impending transition away from fossil fuels.

Drawing insight from the energy humanities, this article explores the significant role played by hydraulic energy, or hydropower, in the process of energy transition during the second half of the nineteenth and early twentieth centuries in Mexico. I will analyze how Mexican intellectuals construed hydraulic energy as the primary basis of industrialization and modernity before oil dependence was established in the country. Mexico had a "later and limited transition from biomass to coal" during the years 1870–1910, and then a "quick transition to oil" between 1900 and the 1920s.[2] Unlike England, where a lengthy period of complete reliance on coal (from the mid-eighteenth century to the 1940s) powered industrialization, Mexico's scant coal deposits were systematically surveyed and exploited only in the 1880s.[3] This is why a combination of coal, biomass, and hydropower was essential to secure the necessary energy supply for industrialization before oil became the energetic basis of Mexican society. In particular, hydraulic energy spearheaded the early efforts of industrializing central Mexico and was still coded as the harbinger of modernization in the 1910s and 1920s. What affects and social imaginaries were associated with hydraulic energy in nineteenth-century Mexico? To what extent was hydropower construed as the basis of an alternative social organization that could compete with oil's predominance in Mexican society? And, finally, how did this historical background allow President López Obrador to resort to hydraulic energy as a kind of ecologically sound, nationalistic way of modernizing the country in the twenty-first century?

In order to approach these questions, I will focus on two time periods that represent two technological and social stages of hydropower exploitation. First, during the 1860s—before coal deposits were surveyed in the country—the kinetic energy of water constituted the power source that could surpass the traditional energetic constraints inherent to agrarian societies in order to fuel energy-intensive industries. During this period, hydropower was directly harnessed by factories that were constructed in riverbanks across the Valley of Mexico. In particular, I will analyze visual and written representations of the textiles factories that were built along the San Ángel River in the 1860s. Second, I will examine how by the 1890s hydraulic energy was being converted into electricity that could be transported to remote places in order to power diverse industrial and domestic activities. Thus, hydroelectricity embodied a utopian promise of progress in the context of the Porfirio Díaz's modernizing regime.[4] Particularly, I will study representations of the hydroelectrical plant constructed in the Necaxa river at the beginning of the twentieth century, which was considered one of the largest plants of its type and provided electricity to Mexico City.

After the Age of Wood

Beginning in the 1860s, the social and ecological consequences of deforestation were a subject of discussion among Mexican intellectual circles.[5] The forests surrounding Mexico City were undergoing a process of uncontrolled exploitation in order to extract wood, the primary source of energy for households and key industries, especially mining. Members of the main scientific organizations—the Mexican Society of Geography and Statistics

(*Sociedad Mexicana de Geografía y Estadística*) and the Mexican Society of Natural History (*Sociedad Mexicana de Historia Natural*)—generally agreed that the ongoing scarcity of wood was compromising the industrialization of the country. Furthermore, some intellectuals such as Ignacio Ramírez and Leopoldo Río de la Loza argued that deforestation had further impacts such as an increase in polluted air that harmed public health, a decline in rainfall and the general change of local climate.[6] In this context, calls were made to adopt fossil fuels—particularly, coal—as a way to prevent excessive deforestation and promote further industrialization.[7] But if the surveying and extraction of coal deposits was a difficult and ambitious plan for the future, tapping into hydraulic resources was already proving successful in the Valley of Mexico and elsewhere.[8] While some machines powered by falling water had existed for centuries, technical innovation had recently produced complex mechanical devices that, along with the standardization of human labor, increased efficiency in the harnessing of energy.

One of the early representations of the promising potential associated with hydraulic energy is José María Velasco's "El Cabrío de San Ángel" (1863), also known as "Fábrica de La Hormiga," which portrays the landscape surrounding the textile factory of La Hormiga located on the San Ángel riverbank in the outskirts of Mexico City (Tizapán). In the early 1860s, Velasco was a young art student at the Academia de San Carlos who was also interested in studying and incorporating natural history into his paintings. In 1865, he decided to formally enroll in courses on zoology and botany at the Escuela de Medicina and participated as scientific illustrator in a research expedition to Puebla. A few years later, Velasco became a member of the Mexican Society of Natural History and published scientific studies in the journal of the Society, *La Naturaleza*.[9] Thus, it is safe to assume that Velasco was aware of the ideas on deforestation and energy transition that circulated in scientific circles and publications by the early 1860s. At the same time, in addition to his interest in natural history, Velasco was also captivated by the possibility of depicting signs of early modernization in Mexico such as factories, railroads, and steel bridges.

"El Cabrío de San Ángel" combines Velasco's interests in modernization and nature in a dramatic landscape. In the right background we discern the imposing walls of the factory and its smoking chimney, whose formidable height nearly reaches the adjacent trees. On top of the factory there is a man, possibly a worker or a watchman, who seems to be surveying the landscape. In the left background the painting depicts the exact location in

FIGURE 6.1. José María Velasco's *El Cabrío de San Ángel* (1863), *Fábrica de La Hormiga*. Photo courtesy of Wikimedia Commons. © Reproduction authorized by the Instituto Nacional de las Bellas Artes y Literatura, 2024.

which the energy stored in the water transforms into kinetic energy as a stream falls down the waterfall of Tizapán. This released energy in motion was harnessed by the nearby factory to power the mechanical processes required for manufacture. In the left foreground we perceive a suspenseful scene: an Indigenous shepherd in traditional clothes is herding his goats away from two ominous threats: on the one hand, the risk of falling off the ledge into the fierce water streams and, on the other hand, a menacing agave that points its sharp edges directly toward them. While the shepherd and his way of life are clearly in danger, the man seated on the roof of the factory is controlling the space with a commanding gaze.[10]

As Fausto Ramírez has suggested, Velasco's painting stages a contrast between "el arcaico (y precario) mundo pastoral y la moderna seguridad que el progreso parecía ofrecer, emblematizada en la industria" (the archaic [and precarious] pastoral world and the modern security that progress seemed to offer, exemplified by industry).[11] It is also a stark contrast between two energy regimes: an agrarian society powered by solar radiation converted through photosynthesis into food for people and cattle, and an industrialized organization fueled by hydraulic energy. The boundary that divides these energy regimes is the river channel through which the water is flowing. Emerging from the river channel and right next to the endangered shepherd, we see a rainbow alongside a huge dark shadow presumably projected by a cloud outside the visual space. This juxtaposition between color and darkness lies at the center of the painting, right in the breach separating the shepherd from the factory, the agrarian from the modern. One could argue that this juxtaposition captures the ambivalent stances and feelings elicited by energy transition in nineteenth-century Mexico and perhaps even in the current times. The shift from one energy system to another often seems to be experienced as a state of suspension full of uncertainty, which may combine the fear toward the demise of ways of life deemed essential and the promise of new and unimagined possibilities.

However, in light of the aforementioned debates on deforestation, I contend that Velasco's painting takes a specific stance on this issue by positing hydropower as an alternative energy basis that solves the problem of the lack of wood and coal while also moving the country forward on the path of modernization and industrialization. Unlike mining, which required the transportation of large amounts of wood to the space of extraction, textile mills could be constructed on riverbanks in order to directly harness renewable energy. Velasco might even be suggesting that hydraulic energy is the material foundation of an ecologically sustainable industrialization that evades the environmental issues caused by deforestation. Even if the painting suggests, through the image of the smoking chimney, what today would be called an ecological footprint, it is indisputable that deforestation—not air or water pollution—was considered the major ecological threat for Mexican conservationists at the time. In this sense, more than an environmental danger, the smoking chimney in this painting was probably coded as a promise of unparalleled social and economic progress.

Moreover, a common thread in the discussion of the time was the argument that traditional agrarian practices—such as communal ownership of the forest commons or burning grasslands to stimulate regrowth—were greatly contributing to deforestation and its negative socioecological outcomes. For example, in the 1850s Río de la Loza maintained that "los indígenas de los pueblos que tienen bosques propios han cortado y cortan a su antojo los árboles, sin sujeción a la ordenanza antigua de bosques y sin observar siquiera el método que aconseja la razón en beneficio propio" (the Indigenous peoples who have their own forests have cut down and keep cutting down the trees at will, without being subject to the old forest ordinance and without even observing the method that reason advises for their own benefit).[12] In this sense, the Indigenous shepherd in Velasco's painting embodied, in the eyes of many Mexican intellectuals, the existence of "primitive," unreasonable or squandering ways of relating to the land and forests. In contrast, the painting implies that the water-powered factory represents a more rational, productive, and ecologically sound system of interacting with nature. This perspective usually implied that, if the country wanted to achieve modernization, Indigenous communities had to assimilate into modern society by necessarily foregoing their "primitive" practices. This is why the Indigenous shepherd in the painting seems to be endangered by his surroundings, especially by the stream that powers the factory.

Velasco's painting stands out among contemporary representations of La Hormiga factory due to its attentive examination of energy issues and its dramatic quality. As a case in point, we can contrast it with a lithograph by Casimiro Castro included in *México y sus alrededores: Colección de vistas monumentales, paisajes y trajes del país*, edited by V. Debray in 1869. Castro's work is a picturesque and idyllic landscape in the spirit of Moritz Rugendas's well known works on Mexico and other Latin American countries. In the middle background, it portrays the Hormiga factory and the waterfall of Tizapán, which transforms into a river stream that in turn culminates in a smaller waterfall in the middle foreground. The factory's chimney, along with the waterfall and the river, form the central axis of the work and divide the landscape into two parts of almost equal proportion. The stream flows peacefully through a highly idealized landscape composed of lavish flora and aesthetically pleasing rocks. Nature is supplemented by the depiction of social stereotypes defined by a few traits such as clothing and occupation. For example, on the left bank of the river we perceive a group of upper-class people wearing sumptuous clothing and enjoying an idle stroll through nature. By the other side of the river there is a woman from a lower social class washing clothes in the river water.

Even though this lithograph depicts the variety of social classes and customs, there is no political antagonism or conflict suggested in this scene of leisure. On the contrary, the different groups seem to coexist peacefully with each other and with the idealized nature. In contrast to the menacing natural features in Velasco's painting, there is nothing threatening or dangerous about the flora or the river in this work. At the same time, unlike Velasco's painting, Castro's lithograph does not suggest a stark contrast between the agrarian and the modern or between energy regimes. Industrialization

FIGURE 6.2. Casimiro Castro's lithograph of La Hormiga factory, included in *México y sus alrededores: Colección de vistas monumentales, paisajes y trajes del país* (1869). Photo courtesy of the New York Public Library.

powered by hydraulic energy does not seem to disrupt traditional agrarian ways of life. The fact that the chimney is not smoking suggests that the factory is not a symbol of modernity, so much as one more of the picturesque components of the scene. In this sense, the couple crossing the bridge over the smaller waterfall is effectively reuniting the two structural parts of the work and suggesting the harmonious merging of the modern and the traditional, the industrial and the agrarian, and the upper and lower social classes. While Velasco's painting employs a high vantage point with the aim of emphasizing a dramatic contrast, this lithograph uses a low vantage point to create the illusion that the spectator is a participant in the idyllic scene.

All in all, Velasco's "El Cabrío de San Ángel" captures perfectly the

Mexican intellectual elites' predicament as they pursued modernization without a viable energy basis that could power it. Uncontrolled deforestation and the scarcity of wood were deemed the main socioecological problems jeopardizing the modernization of the country in the eyes of 1860s Mexican conservationists. While the intensive use of fossil fuels was an unreachable goal at the time, the construction of textile mills fueled by hydraulic energy provided a model of an ecologically sound industrialization. The social and economic potential associated with hydraulic energy intensified by the end of the nineteenth century, when technological innovations allowed the conversion of water's kinetic energy into electricity.

The Promise of Hydroelectricity

In 1918, conservationist engineer Miguel Ángel Quevedo delivered a speech at the Mexican Society of Geography and Statistics in which he talked about the crucial role played by hydroelectricity in the last three decades. Quevedo started by assessing the strengths and challenges of each kind of energy in regard to fulfilling the energy requirements of modern capitalist societies. According to him, it is evident that exclusive reliance on wood could not have provided the material basis for the industrial revolution that transformed the Western world: "no bastarían los bosques aun los de más intensa vegetación y buena calidad, por muy bien administrados que se les suponga, para proveer las necesidades del consumo en fuerza motriz y calefacción de las sociedades modernas" (Forests, even those with the most intense vegetation and good quality, would not be enough, no matter how well managed they might be, to supply the amount of driving force and heat modern societies consume).[13] Coal fueled England's "supremacía naval e industrial" (naval and industrial supremacy) during the nineteenth century, but this type of energy has inherent limits because "aunque rinde más que la leña, (el carbón) es todavía muy bromoso y poco aprovechable en su energía" (Although it yields more than wood, (coal) is still very cumbersome and little exploitable in its energy).[14] At the same time, hydrocarbons such as oil certainly yield high performance, but are nevertheless far from being the definitive response to society's requirements. As Quevedo puts it, "cuál lejos de poder ser ese recurso natural de los hidrocarburos la panacea dinámica o calorífica que venga a proveer todas las necesidades mundiales. Por muy abundantes que se suponen sus reservas subterráneas, éstas son de pronto agotamiento en el indefinido y cada vez mayor consumo de energía y calor que hace la humanidad" (how far from being the dynamic or calorific panacea that comes to supply all the world's needs the natural resource of hydrocarbons is. However abundant its underground reserves are supposed to be, they face a fast depletion in humanity's indefinite and increasing consumption of energy and heat).[15] In addition to being nonrenewable, oil has the problem of being "insalubre" (unhealthy) because it produces "alteración de la atmósfera por los productos mismos de la combustión y el consumo del oxígeno" (alteration of the atmosphere by the products of combustion and oxygen consumption).[16]

In opposition to all those energy sources (wood, coal and oil), hydraulic energy is the only one that is "inagotable" (inexhaustible) and provides a "manantial de energía siempre espontáneamente renovable" (source of always spontaneously renewable energy), since it depends on recurring natural processes such as the water cycle.[17] Quevedo goes on by reviewing the history of harnessing hydraulic energy in Mexico and maintains that, as I have shown, "en nuestra República la industria fabril de hilados y tejidos encontró en ella (energía hidráulica) poderoso y muy económico elemento desde sus principios" (In our Republic the manufacturing industry of textiles and fabrics found in it [hydraulic energy] a powerful and very economical element from its beginnings).[18] However, in the 1890s the shortage of wood and coal provoked a crisis that hampered the industrialization of the country. According to Quevedo, hydroelectricity effectively saved Mexico from this dire crisis and powered its modernization during the Porfirio Díaz administration. As he points out, "Habría sido imposible el desarrollo de las industrias extractivas y fabriles de nuestro país en general, si al acentuarse dicha crisis por falta de combustible económico en aquellos años de 1890 a 1895, no hubiera venido tan oportunamente la magnífica aplicación de la transformación de la energía hidráulica en energía eléctrica y su transporte aéreo a grandes distancias, con perfectos requisitos de seguridad, a proporcionar medio económico para salvar nuestra grave crisis industrial" (The development of the extractive and manufacturing industries of our country in general would have been impossible, if when said crisis worsened due to lack of affordable fuel in those years from 1890 to 1895, the magnificent application of the transformation of hydraulic energy into electrical energy and its air transport over long distances, with perfect safety requirements, had not come so opportunely to provide an economic means to save our serious industrial crisis).[19]

Quevedo states that in 1897 there were two important hydroelectrical plants in Mexico, one in the Río Grande Santiago (Jalisco) and the other in the Río Blanco (Veracruz), but in the next two decades several other plants were built throughout Mexico with remarkable socioeconomic consequences.[20] Quevedo's assessment of hydroelectricity's historical impact on the country was shared by many intellectuals and statesmen of the time, including Porfirio Díaz himself.[21] For example, John Hubert Cornyn, a professor of English at the Universidad Nacional de México, remarks in his 1910 book *Díaz y México*: "La historia del progreso industrial de México durante la última década (1900–1910), no es sino la historia del desarrollo de su fuerza hidráulica" (The history of industrial progress in Mexico during the last decade [1900–1910], is nothing but the history of the development of its hydraulic power).[22] He then goes on by outlining the magnificent industrialization made possible by hydroelectric energy: "La provisión de fuerza, a la tercera parte del precio que antes costaba, ha hecho posible la operación de centenares de fábricas, grandes y pequeñas; ha triplicado la extensión de los trabajos mineros; ha convertido la luz eléctrica en una comodidad cuando antes era un lujo; ha aumentado las líneas urbanas y suburbanas de tranvías; y en fin, ha revolucionado por completo las condiciones industriales" (The supply of power, at a third of its former price, has made possible the operation of hundreds of factories, large and small; the extension of mining sites has tripled; it has turned electric light into a convenience when it used to be a luxury; it has increased urban and suburban tram lines; and finally, it has completely revolutionized industrial conditions).[23] In this sense, both Cornyn and Quevedo assume that the modernization brought about by the Díaz administration would have been impossible without the energy basis provided by hydroelectricity.

Among the hydroelectrical plants built in Mexico at the turn of the century, the Necaxa plant—constructed by the Mexican Light and Power Company between 1903 and 1905 in the state of Puebla—was unanimously considered the most outstanding and impressive. It was often hailed as a technological marvel that embodied modernity and allowed Mexico to compete with the most advanced industrial nations: "La planta eléctrica de Necaxa, donde se produce la energía, puede considerarse como de lo más moderno que existe hoy en el mundo en su género" (The Necaxa power plant, where energy is produced, can be considered one of the most modern of its kind in the world today).[24] As Quevedo argues, the Necaxa facilities "son sin duda, por las diversas circunstancias técnicas y problemas que hubo que resolver, de las más importantes y meritorias del país y aun del mundo; ellas tienen capacidad para producir una fuerza efectiva de sesenta mil caballos, con cuyo importante contingente se ha venido a proveer en pocos años con abundancia a las necesidades de energía para alumbrado y fuerza motriz de la Ciudad de México y demás poblaciones del Distrito Federal y otras regiones" (are undoubtedly, given the various technical circumstances and problems that had to be solved, among the most important and meritorious in the country and even the world; they have the capacity to produce as much power as 60,000 horses, and in a few years have abundantly met the energy needs for lighting and motor power of Mexico City and other populations of the Federal District and other regions).[25] Certainly, in contrast to the relatively simple hydraulic system installed at La Hormiga factory in the 1860s, the Necaxa hydroelectric plant required colossal work in order to feed the water from the Necaxa river—and other nearby rivers such as the Tenango—to turbines and electrical generators by way of a series of tunnels, pipe lines and dams. The electricity was then transported 170 miles on cables to power streetlamps, trolley cars, and factories in Mexico City, as well as the mining region of El Oro.[26] Quevedo praises the engineers who envisioned and worked on Necaxa as "hombres sumamente competentes en lo técnico y en lo económico" (highly competent men in technical and economic matters) who would not have invested such a large amount of capital if they believed that other energy sources such as oil or coal would outperform hydroelectricity.[27]

Necaxa's technological magnificence was often construed as mastery over an equally magnificent and sublime landscape. Since the 1850s, the natural surroundings of the Necaxa river captured the imagination of travelers, artists and scientists who were especially attracted to the enormous and stunning waterfall that became known as waterfall of Huachinango or Necaxa.[28] One of the early accounts of this landscape was a brief article by José Justo Gómez de la Cortina, who described the waterfall as "(uno de) los objetos más grandiosos y magníficos con que la naturaleza ha querido enriquecer a la República Mexicana" ([one of] the greatest and most magnificent objects with which nature has enriched the Mexican Republic).[29] He lamented that, even though Mexican intellectuals frequently admire distant waterfalls such as Niagara Falls, the waterfall of Necaxa remained mostly unexplored and overlooked. According to Gómez de la Cortina, the Necaxa waterfall stands out for two reasons: first, "la frondosidad, variedad y riqueza de sus terrenos" (the lushness, variety and richness of its land) creates an unparalleled landscape in which the waterfall constitutes only one of the most salient features.[30] Additionally, the height of the Necaxa waterfall is triple that of Niagara Falls, producing an exceptional scene where the water reaches the floor: "es indescriptible la fuerza con que chocan, se agitan, hierven y se levantan enormes volúmenes y remolinos de agua conmovidos, rechazados y trastornados en todas direcciones" (the force with which [the stream] collides, shakes, boils and gives rise to enormous volumes and whirlpools of water, stirred, rejected and disordered in all directions, is indescribable).[31] Like other romantic travelers of the time who encountered waterfalls, Gómez de la Cortina draws on a sense of sublimity by emphasizing the vastness that overtakes human notions of reason or measurement and elicits a strong emotion in the viewer.[32]

In the 1860s, the Secretaría de Fomento organized the first scientific expedition to the area surrounding the Necaxa waterfall. Composed of engineers Ramón Almaraz and Guillermo Hay, geographer Antonio García Cubas, and artists José María Velasco and Luis Coto, this expedition was formed to study recently discovered archeological remains and consider the plausibility of establishing settlements. Most of the accounts the expedition's members produced relied upon the same set of cultural assumptions as nurtured Gómez de la Cortina's early description. For example, García Cubas refers to the waterfall of Necaxa as "el prodigioso salto" (the prodigious fall), the magnificence of which is intensified by its natural setting.[33] As he points out, "Si apartábamos la vista de aquel espectáculo sorprendente, observábamos, cualquiera que fuese el punto a que la dirigiéramos, otros tan dignos de admiración, pues en aquellos lugares reinan por completo las armonías providenciales; ya sean elevadas y fértiles praderas ilimitadas por boscosas eminencias; ya grietas profundas y estrechas cañadas como la de que tratamos y en la que, recobrando el agua su normal movimiento, se desliza, ora en rompientes por el centro de su cauce, ora tranquila por las riberas bordadas de corpulentos árboles y preciosas flores" (If we looked away from that surprising spectacle, we would observe, whatever the point to which we directed our gaze, other ones so worthy of admiration, because in those places providential harmonies reign completely; whether they are high and fertile meadows bounded by wooded hills; or deeps cracks and narrow ravines like the one we are describing and through which the water slides, regaining its normal movement, now breaking through the center of its channel, now quietly along the embroidered banks of stout trees and precious flowers).[34] García Cubas highlights the intense variety and sumptuosity of natural features from geological formations to lush vegetation to streams.

FIGURE 6.3. José María Velasco's *Cascada de Necaxa*. From *Memoria acerca de los terrenos de Metlaltoyuca* (1866), edited by Ramón Almaraz. © Reproduction authorized by the Instituto Nacional de Bellas Artes y Literatura, 2024.

Similarly, Velasco created a drawing that accentuates the sublime vastness of the landscape by providing a wide frame showing the Necaxa waterfall surrounded by immense mountains and deep canyons.

Thus, by the time the Necaxa plant was built the social construction of this sublime landscape was firm enough to shape social imaginaries concerning the hydroelectrical plant. It was assumed that only an extremely powerful technology and titanic engineering could achieve the grand mission of conquering and dominating the sublimity of the waterfall in order to put it to work on behalf of modernization. American journalist Wallace Thompson, who supported the Díaz administration as a news editor of the *Mexican Herald* at Mexico City, emphasized how the hydroelectric plant practically stopped the water falling down the ledge, establishing a stark contrast between the original waterfall and what was left after the construction of the plant:

> Hace un año la maravillosa caída inferior de Necaxa era una de las más hermosas cataratas imaginables. . . . Las blancas nubes de átomos húmedos producían uno de los más hermosos efectos que pudieran imaginarse y el rugido del agua resonaba por la barranca como el bramido de los laboratorios de la Naturaleza. Hoy día la caída es apenas una delgada corriente donde antes se desprendía en tres poderosas secciones que se unían para formar una gran cortina nebulosa antes de llegar al pie de las caídas.

> A year ago, the wonderful lower fall of Necaxa was one of the most beautiful waterfalls imaginable. . . . The white clouds of wet atoms produced one of the most beautiful effects that could be imagined and the roar of the water echoed through the ravine like the roar of Nature's laboratories. Today the fall is just a thin stream where it used to break off into three powerful sections that joined to form a great hazy curtain before reaching the foot of the falls.

Thompson goes further and suggests that the waterfall had to surrender its sublime greatness and even *transfer* it to the hydroelectrical installations, which in turn would power the wonders of modernization: "Mañana el Valle de México, los Estados de México y Puebla e Hidalgo y . . . la misma capital de la República, se animarán con el ruido de rápidos motores y el murmullo de las máquinas que invadirán el espacio lo mismo que el de la gran catarata que ha dado su hermosura y su vida para que estas cosas existan" (Tomorrow the Valley of Mexico, the states of Mexico and Puebla and Hidalgo and . . . the very capital of the Republic, will be animated by the noise of fast engines and the murmur of machines that will invade space just as the great waterfall that has given its beauty and its life so that these things exist).[35] According to Thompson, the Necaxa waterfall had to be sacrificed in the name of modernity, but its sublime condition remains untouched in the grandiose hydroelectrical project, which now embodies a kind of "technological sublime."[36]

Thompson's suggestion is consistent with the Porfirio Díaz administration's stance on the Necaxa plant, as demonstrated in a photograph depicting the official inauguration of the hydroelectrical facilities in 1905. During

FIGURE 6.4. Inauguration of the Necaxa Hydropower Plant, 1905. Photo courtesy of Siemens Historical Institute.

the Díaz administration, the inauguration of massive infrastructural constructions such as railroads or buildings became highly ritualized ceremonies that sought to reinforce and publicize the administration's slogan: "orden y progreso."[37] Carefully staged photographs were usually taken and then published in newspapers and magazines. In this case, we see in the center of the photograph an official delegation that includes engineers, state officials, and Díaz himself in a prominent position looking directly at the camera. The group of people is standing on a steel platform hanging over the cliff, which appears right between a transmission line tower to the left and a waterfall to the right. While the electrical tower stands so tall that it easily extends beyond the visual space, the stream falls from such a high altitude that it also disappears from view. By thus emphasizing their unthinkable magnitude, the composition of the photograph suggests that both the transmission line tower and the waterfall are sublime objects that surpass standard measurements. Díaz and his cohort of engineers and state officials play an intermediary role between the magnificent objects: their technical knowledge and expertise has allowed them to tame the natural sublime in order to create the technological sublime. The fact that they are hanging over the cliff suggests that this intermediary role is a daring endeavor that should be undertaken only by knowledgeable and powerful people.

This photograph may also shed light on the ways in which the Díaz administration carried out the Necaxa project vis-à-vis the Indigenous population that lived in the area. In the left of the photograph there are a few Indigenous men who are watching Díaz and the official delegation. Their location outside of the central steel platform, as well as their traditional clothing and sombreros—which contrast with the Western-style hats and fancy suits worn by Díaz and the group of distinguished people—clearly imply that these onlookers do not participate in the technological progress associated with the Necaxa plant. In the eyes of the intellectual elite of the time, these Indigenous persons represented one of the main factors that hindered the program of modernization and technological development espoused by the Díaz administration. Just as Velasco's painting of La Hormiga factory suggests the extinction of the Indigenous shepherd, this photograph also displays the residual and subaltern position assigned to Indigenous populations by intellectual and political elites. In fact, this became evident during the construction of the hydroelectrical plant because it required the forced displacement of the entire town of Necaxa in order to build a dam in its place.[38] In addition to involuntary resettlement, building the dam, pipe lines, and related hydroelectrical infrastructure almost certainly had negative social and ecological consequences like those documented at other sites, where the alteration of river streams endangers organisms' habitats and patterns of mobility, ultimately depriving surrounding communities of access to vital natural resources (irrigation water, timber, fish).[39] From the perspective of local and national elites, however, the demise of the town and ecosystem of Necaxa were construed as the small, unavoidable price to be paid in order to reach the goal of socioeconomic and technological modernization.

The Necaxa plant reached its peak of production in the first three decades of the twentieth century. During the same period Mexico became one of the largest exporters of oil in the world and increased domestic consumption so as to be largely dependent on oil and its derivatives by the 1930s.[40] Since then, the ever-growing extraction of oil—not hydraulic energy anymore—emerged as the major promise of Mexico's modernization and prosperity in the future. Finally, Necaxa and other hydroelectrical plants were nationalized and taken over by the state in 1960 during the Adolfo López

Mateos administration. But perhaps some of the affects and social imaginaries attached to hydroelectricity still remain active to this day, as seen in president López Obrador's plan to launch an ecologically sound modernization based on the optimization of state-owned hydroelectrical plants.

Closing Remarks

This article has shown how hydraulic energy was coded as a harbinger of modernization in Mexico during two distinct time periods. In the 1860s, as epitomized by Velasco's "El Cabrío de San Ángel," hydropower was presented as a way of bolstering industrialization in the face of the lack of wood and coal that jeopardized Mexico's industrialization. Moreover, hydraulic energy was posited as an optimized and efficient method of exploitation—in contrast with "primitive" ways of extracting natural resources—that could contribute to hindering deforestation's potential impact on the environment. By the end of the nineteenth century, the positive connotations already associated with hydropower increased and intensified due to the construction of hydroelectrical projects in Mexico such as the Necaxa plant. According to conservationists like Quevedo, hydroelectricity was still considered a cleaner, more available, and efficient source of energy than hydrocarbons, which were being surveyed and exploited in the country with great success. In the view of Porfirio Díaz and aligned intellectuals, the massive and magnificent character of the Necaxa plant elicited a sense of technological sublime that ultimately legitimized the control and optimization of nature. In this sense, it is clear that natural ecosystems—along with Indigenous peoples—were forced to bear the costs of the modernization brought about by the Necaxa plant and the developmental agenda of the Díaz administration.

The role played by hydropower in the energy transition to fossil fuels allows us to gain insight into how to understand and approach the necessary transition away from fossil fuels. In nineteenth-century Mexico, hydropower was predominantly considered an ecologically sound energy source, but it never actually represented an alternative social organization to fossil fuel-based capitalism. Even when Mexican intellectuals advocated the ecological advantages of hydropower, their ultimate and authentic ambition was to put Mexico on the path of modernization spearheaded by advanced nations such as Britain and the US. Ultimately, the development of hydraulic energy was predicated upon the same core values that under-pinned the expansion of fossil-fuel use: the need for a self-sustaining, end-less socioeconomic growth based on a cheap, supposedly unlimited, power-ful source of energy. In the end, the exploitation of both hydropower and fossil fuels relied on a shared framing of the earth as a standing reserve of energy and resources awaiting extraction. This shows us how far-reaching and deeply ingrained the influence of fossil fuels was during the nineteenth century—and still is to this day—in the general outlook of modern soci-ety, including its socioeconomic regime and expectations, ways of relating to the earth, cultural representations, and so on. Understanding how pro-foundly seated fossil fuel use is obliges us to recognize that the current task of imagining non-oil-based forms of social organization cannot restrict it-self—as President López Obrador seems to think—to the idea of incorpo-rating clean, renewable energies in otherwise unchanged social structures. On the contrary, this challenging task requires correcting the oversight of nineteenth century intellectuals by calling into question the fundamental notions of what "development" and "modernity" stand for, with the aim of opening up new possibilities and ways of living for the future.

NOTES

1. Szeman and Boyer, *Energy Humanities*, 2–3.
2. Vergara, "Energy, Environment, and Society," 3. See also Vergara's book-length study of this energy transition, *Fueling Mexico*.
3. Vega y Ortega Baez and Serrano Jurárez, "Los estudios sobre el carbón," 64.
4. In this vein, Montaño's *Electrifying Mexico* examines the emergence of "dreams and visions of an electrified future" (16) during the Porfirio Díaz administration.
5. Vergara, *Fueling Mexico*, 84–87.
6. Urquiza García, "Historia ambiental y problemas," xxi–xxiii.
7. Vergara, "How Coal Kept My Valley Green," 4–5.
8. Vergara, *Fueling Mexico*, 52–53.
9. Trabulse, *José María Velasco*, 119–217.
10. Ramírez, *José María Velasco: Homenaje*, 27.
11. Ramírez, *José María Velasco: Pintor de paisajes*, 25.
12. Río de la Loza, *Vivir para conservar*, 3.
13. Quevedo, *Vivir para conservar*, 266.
14. Quevedo, *Vivir para conservar*, 266, 267.
15. Quevedo, *Vivir para conservar*, 268.
16. Quevedo, *Vivir para conservar*, 268.
17. Quevedo, *Vivir para conservar*, 268, 269.
18. Quevedo, *Vivir para conservar*, 269.
19. Quevedo, *Vivir para conservar*, 270.

20. Vergara, *Fueling Mexico*, 120.
21. For an examination of Díaz's views on electricity, see Montaño's *Electrifying Mexico*, 8–11.
22. Cornyn, *Díaz y México*, 214.
23. Cornyn, *Díaz y México*, 214.
24. Cornyn, *Díaz y México*, 216.
25. Quevedo, "El porvenir del carbón," 274.
26. For an analysis of cultural representations of Mexico City as an "electric city," see Montaño's *Electrifying Mexico*.
27. Quevedo, "El porvenir del carbón," 277.
28. Checa-Artasu, Sunyer Martín, Francisco Coello, *La electrificación y el territorio*, 8–10.
29. Gómez de la Cortina, "Cascada de Huachinango," 155.
30. Gómez de la Cortina, "Cascada de Huachinango," 155.
31. Gómez de la Cortina, "Cascada de Huachinango," 156.
32. Hudson, "Waterfalls and the Romantic Traveler," 41.
33. García Cubas, *El libro de mis recuerdos*, 578.
34. García Cubas, *El libro de mis recuerdos*, 585.
35. Thompson, "Fuerza hidráulica de Necaxa," 3, quoted in Checa-Artasu, Sunyer Martín, and Francisco Coello, "De lo indispensable," 12.
36. Nye, *American Technological Sublime*, xiii–xv.
37. Montaño, *Electrifying Mexico*, 10.
38. Peña Guzmán, "La hidroeléctrica de Necaxa," 53.
39. McCully, *Silenced Rivers*, 29–100; Miller, *An Environmental History*, 157–65.
40. Uhthoff López, "La industria del petróleo," 33, 9–13.

BIBLIOGRAPHY

Checa-Artasu, Martín, Pere Sunyer Martín, José Francisco Coello. "De lo indispensable a lo incómodo: El complejo hidroeléctrico de Necaxa (México) (1895–2016) como paisaje cultural." In *La electrificación y el territorio: Historia y futuro*, edited by Horacio Capel Sáez, Miriam Zaar, Magno Vasconcelos P. Junior, 1–35. Universidad de Barcelona/Geocrítica, 2017.

Cornyn, John Hubert. *Díaz y México*. Imprenta Lacaud, 1910.

García Cubas, Antonio. *El libro de mis recuerdos: Narraciones históricas, anecdóticas y de costumbres mexicanas anteriores al actual estado social*. Imprenta de Arturo García Cubas, Hermanos Sucesores, 1904.

Gómez de la Cortina, José Justo. "Cascada de Huachinango." *Boletín de la Sociedad de Geografía y Estadística de la República Mexicana* (1853): 155–56.

Hudson, Brian J. "Waterfalls and the Romantic Traveler." In *Appreciating Physical Landscapes: Three Hundred Years of Geotourism*, edited by T. A. Hose, 41–57. Geological Society Special Publications, 2016.

McCully, Patrick. *Silenced Rivers: The Ecology and Politics of Large Dams*. Zed Books, 2001.

Miller, Shawn William. *An Environmental History of Latin America*. Cambridge University Press, 2007.

Montaño, Diana. *Electrifying Mexico: Technology and the Transformation of a Modern City*. University of Texas Press, 2021.

Nye, David. *American Technological Sublime*. MIT Press, 1994.

Peña Guzmán, Celina. "La hidroeléctrica de Necaxa y la Mexican Light and Power Co., patrimonio industrial en riesgo." *Revista Labor and Engenho* 5 no. 2 (2011): 45–65.

Quevedo, Miguel Ángel. "El porvenir del carbón blanco en la República Mexicana." In *Vivir para conservar: Tres momentos del pensamiento ambiental mexicano: Antología*, edited by Juan Humberto Urquiza, 265–84. UNAM, 2018.

Ramírez, Fausto. "Acotaciones iconográficas a la evolución de episodios y localidades en los paisajes de José María Velasco." In *José María Velasco: Homenaje*, 15–85. UNAM, 1989.

Ramírez, Fausto. *José María Velasco: Pintor de paisajes*. FCE, UNAM, 2017.

Río de la Loza, Leopoldo. "Tala de bosques y exportación de maderas." In *Vivir para conservar: Tres momentos del pensamiento ambiental mexicano: Antología*, edited by Juan Humberto Urquiza García, 3–11. UNAM, 2018.

Szeman, Imre, and Dominic Boyer. "Introduction: On the Energy Humanities." In *Energy Humanities: An Anthology*, edited by Imre Szeman and Dominic Boyer, 1–13. John Hopkins University Press, 2017.

Thompson, Wallace. "Fuerza hidráulica de Necaxa." *Modern Mexico* 21, no. 2 (1906).

Trabulse, Elías. *José María Velasco: Un paisaje de la ciencia en México*. Secretaría de Educación del Estado de México, 2012.

Uhthoff López, Luz María. "La industria del petróleo en México, 1911–1938: Del auge exportador al abastecimiento del mercado interno. Una aproximación a su studio." *América Latina en la historia económica* 33 (January–June 2010): 7–30.

Urquiza García, Juan Humberto. "Historia ambiental y problemas ecológicos contemporáneos." In *Vivir para conservar: Tres momentos del pensamiento ambiental mexicano: Antología*, edited by Juan Humberto Urquiza García, vii–lxxxix. UNAM, 2018.

Vega y Ortega Baez, Rodrigo, and José Daniel Serrano Jurárez. "Los estudios sobre el carbón en la revistas minero-mineralógicas de la Ciudad de México, 1870–1879." *Estudios de historia moderna y contemporánea de México* 54 (2017): 62–75.

Vergara, Germán. "Energy, Environment, and Society in the Basin of Mexico until the Nineteenth Century." In *Mexico in Focus: Political, Environmental and Social Issues*, edited by José Galindo, 1–25. Nova Science Publishers, 2014.

Vergara, Germán. *Fueling Mexico: Energy and Environment, 1850–1950*. Cambridge University Press, 2021.

Vergara, Germán. "How Coal Kept My Valley Green: Forest Conservation, State Intervention, and the Transition to Fossil Fuels in Mexico," *Environmental History* 23, no.1 (January 2018): 82–105.

CHAPTER 7

That Mysterious *Something*

Nature, Mystery, and Animism in W. H. Hudson's Early Writings

LESLEY WYLIE

William Henry Hudson (1841–1922) was born near Quilmes, Buenos Aires, to North American parents and lived in Argentina until his departure for England in 1874. Although he never returned to South America, these years defined both his understanding of nature and his distinctive written style. A contemporary and acquaintance, Joseph Conrad, said that Hudson's writing "was like the grass that the good God made to grow and when it was there you could not tell how it came."[1] This oft-quoted simile not only firmly situates Hudson's writing in its pampas environs, where grass (particularly *Gynerium argenteum*, "the stately pampa grass") is the preeminent vegetation, but draws out its organic, spontaneous nature as well as its inherent mysteriousness—a quality that, as Jason Wilson notes, Conrad saw as emblematic of Hudson's life as well as his art.[2]

Hudson's collected works, published by Dent in 1923, run to twenty-four volumes and embody the "fascinating mysteriousness" that Conrad discerned in the Argentine author.[3] His literary output was eclectic, consisting of novels, short stories, poems, memoir, natural history, and rural tales, many collected on his rambles in the south of England. He wrote of

Argentine gauchos, and English shepherds, of lost and waning traditions, of political upheaval, human endurance, vegetarianism, the supernatural and—especially—of encounters with wild animals, often birds. Much criticism on Hudson has focused on his writing's "intermediary position" between aesthetics and science.[4] Felipe Arocena regards Hudson's work as "nourished by the opposition between Romanticism and Enlightenment," with the Argentine taking on the role of what Henry Seidel Canby called "Poet Scientist": a man "able to observe and to express" and, thereby, turn "matter into spirit."[5]

This chapter considers anew the persistent opposition between scientific and non-scientific thinking in Hudson, specifically in his earliest naturalist writings on South America: *Argentine Ornithology* (2 vols, 1888–89; revised in 1920 as *Birds of La Plata*), *The Naturalist in La Plata* (1892), and *Idle Days in Patagonia* (1893), an account of a year spent "idling" in the south of Argentina following a shooting accident that rendered him immobile. I open the chapter by examining Hudson's distrust of scientific rationalism and exploring the alternative modes of knowledge advanced in his early writings, including the place of "mystery" and the supernatural in any understanding of nature. In part two I consider Hudson's animistic imagination, which he defines as the "mind's projection of itself into nature, its attribution of its own sentient life and intelligence to all things."[6] Sara Castro-Klarén has tellingly described Hudson's approach to nature as closer to shamanism than the European classificatory tradition.[7] This chapter extends this reading to examine how Hudson's early South American texts redefine the relationship between humans and nature, regarding nonhuman life not as an insentient object of knowledge, but a mysterious, animate realm.

"The Supernatural in All Natural Things": *Science and Myth in Hudson's Early Writings*

Hudson's writing is punctuated with references to science and, especially, to scientists, many of whom had, like him, spent their formative years in South America. He speaks of Alexander von Humboldt, Charles Darwin, and Alfred Russel Wallace; of the Spanish naturalist Félix de Azara, who made important contributions to Argentine ornithology; of the French botanist Claudio Gay, whose time in Chile is recorded in his influential *Historia física y política de Chile*; and of Gilbert White, the British field naturalist upon whom

Hudson modelled himself, calling La Plata, "my 'parish of Selborne'" (*NP*, 5). Hudson was not a trained scientist, but, as Wilson characterizes him, an "amateur and an autodidact," who learned through observation and—evidently—wide reading.[8] Although in his writings he refers frequently to scientific theories such as evolution and mimicry, he also stresses the importance of personal experience of nature—a nature observed up-close, that one can touch, taste, and smell. His frequent quibbles with Darwin were usually grounded on the objection that, as a non-native of South America, the famous author of *The Origin of Species* could not be as knowledgeable about its plants and animals as someone who grew up among them. This resulted, for instance, in an infamous letter, sent by Hudson to the Zoological Society of London in 1870, which challenged Darwin's claim that woodpeckers had adapted to the treeless environment of the Pampas.[9]

Wilson has discussed Hudson's uneasy relationship with science and particularly with his chief antagonist, Darwin.[10] Although Hudson was an erudite and, as Arocena notes, systematic naturalist, his approach to writing about nature resists the empiricism and rationality of much late-nineteenth-century science, especially the belief that nature was entirely knowable and reducible to a series of facts.[11] It is notable that, when he came in 1920 to revise his first naturalist work—the two volume *Argentine Ornithology*, compiled in collaboration with the then "chief authority . . . on South American Ornithology," Dr Philip Lutley Sclater, Hudson excised all of Sclater's classificatory material, leaving only his own "account of the birds' habits."[12] The narrative that remains is rich and engaging. Hudson provides the scientific nomenclature for birds, but not without the odd jibe at the scientific establishment, as, for example, when discussing the Bellicose Tyrant or *Tyrannus melancholicus*, whose name "does not seem altogether inappropriate: that is the most that can be said of any specific name invented by science" (*BP*, 156–57). Alongside scientific designations, Hudson lists Amerindian and Spanish names, which often reveal more about the birds' habits and local associations. Of the *Furnarius rufus*, for instance, Hudson explains:

> It ranges throughout the Argentine Republic to Bahia Blanca in the south, and is usually named *Hornero* or *Casera* (Oven-bird or Housekeeper). . . . In Paraguay and Corrientes it is *Alonzo Garcia* or else *Alonzito*, the affectionate diminutive. Azara, that sensible naturalist, losing his mind for a moment, solemnly says that he can give no reason for such a name! He might have found the reason in his own country in Europe, where as a boy he knew

the wild bird life and where a bird which inspires affectionate admiration in the country people is sometimes called by a human name. . . . The *Alonzo Garcia* is specially favoured in having both a Christian and a surname. I have often been assured by natives that the *Hornero* is a religious bird and always suspends his labours on a Sunday and on all holy days. (*BP*, 165)

Here it is the scientist—a figure normally credited with rationality and sense—who is mocked for a lack of "reason" bordering on madness, whilst the locals' extravagant beliefs are repeated with a deadpan earnestness suggestive of respect if not credulity. Hudson ends his account of the *hornero* a few pages later with an anecdote about how an Argentine neighbor of his (a so-called "old native"; *BP*, 169) witnessed a pair building a nest on the "sepulchre" of one of their late mates. Hudson concludes the chapter on a deepening note of superstition: "it was not strange that, after witnessing the entombment of the one that died, he was more convinced than ever that the little House-builders are 'pious birds'" (*BP*, 169).

The tendency in Hudson to present local knowledge about the natural world as of equal or greater value to scientific interpretation surfaces throughout his early naturalist writings. In *The Naturalist in La Plata* Hudson dedicates a chapter to the puma, in which he examines the enigma surrounding the cat's well-attested docility toward humans. Having dismissed Azara's earlier accounts of the subject as "not a full statement of the facts" (*NP*, 32), and criticizing "all travellers and naturalists" (*NP*, 37) for hitherto misinterpreting the puma's docility as cowardice, Hudson goes on to appeal to a range of local informants (a "native," *NP*, 34; "a person who had spent most of his life on the pampas," *NP*, 42; "a sheep-farming Scotchman," *NP*, 44; and "many gauchos," *NP*, 43) to get to the bottom of the "mysterious gentle instinct of this ungentle species" (*NP*, 37). Putting local oral histories into dialogue with written accounts by Azara, Gay, and Vice-Admiral Robert Fitzroy (captain of the Beagle when Darwin made his famous voyage to Tierra del Fuego) among others, Hudson develops a narrative that subtly blurs the lines between fact and fiction. The generic instability of Hudson's chapter on the puma—which includes travellers' tales, gaucho superstitions, natural history, and eyewitness accounts of the animal—might be seen as characteristically Creole, following on from classic composite works such as Domingo F. Sarmiento's 1845 *Facundo*, and anticipating the stylistic heterogeneity of the *novela de la tierra* tradition of the 1920s and 30s. Hudson's investigations into the puma culminate in his inclusion of an "incident" which had taken place in the district of Saladillo a few years before. Hudson's "first hand" (*NP*, 49) informant relates how, during a hunt, one of the participants disappeared. Discovered the following morning with a broken leg, the hunter tells how, at nightfall, a puma had "sat near him, but did not seem to notice him." A few hours later, "he heard the deep roar of a jaguar, and gave himself up for lost" (*NP*, 49). What follows—testimony of the puma's "extraordinary" (*NP*, 49) defense of the hunter—is presented by Hudson as not only indubitable ("there was really no room for doubt," *NP*, 49), but as evidence of a broader mystery surrounding the relationship between humans and pumas, which in this case is more readily accommodated by local lore than scientific knowledge. Jens Andermann has noted the importance of storytelling in Hudson's fictional writing as a means to breach the epistemological gap that science introduces between the human and the nonhuman.[13] Hudson also embraces local modes of knowledge in his nonfiction writing, as alternative and often superior forms of meaning-making that allow for metaphysical, non-rational, and religious reflection. Such a move aligns Hudson not only with the European Romantic tradition, as I will explore further, but fin de siècle South American anti-Positivists, such as Uruguayan José Enrique Rodó (1871–1917), who also resisted scientific dogma in favor of spiritual and aesthetic approaches to nature.[14]

Throughout his early naturalist writings Hudson accepts—indeed, often advances—local, non-scientific thinking about nature. Hudson insists that the naturalist must "cast aside his books . . . and go directly to nature" (*NP*, 283), advocating a kind of "pure" encounter with the natural world, and repeating the Rousseauian tenet that learning is, paradoxically, inimical to knowledge.[15] This aligns Hudson's thinking about nature firmly with the Romantics and North American Transcendentalists such as Ralph Waldo Emerson and Henry David Thoreau.[16] In *Idle Days in Patagonia* Hudson observes: "Doubtless man is naturally scientific, and finds out why things are not what they seem, and gets to the bottom of all mysteries; but his older, deeper, primitive, still persistent nature is non-scientific and mythical" (33). Throughout *Birds of La Plata* it is the nonscientific and mythical that Hudson persistently appeals to in his presentation of the bird-life he so keenly observed from a young age. He records the "miraculous mocking-powers" (*BP*, 10) of the White-Banded Mocking-Bird, comparing its call to the "ethereal rapturous character" (*BP*, 11) of lark-song. He describes an "exquisitely beautiful" hummingbird as a "miracle of energy" (*BP*, 202), and notes how the "crimson bosoms" of Military Starlings "seem to glow with a strange splendour" (*BP*, 108). *The Naturalist in La Plata* and *Idle Days in Patagonia* accrue more examples of the mysteries of nature, such as the "extraordinary" (*NP*, 49) account of the puma, discussed earlier. Sometimes these mysteries are rather mundane ("Why, or how, animals came to be possessed of the power of emitting pestiferous odours is a mystery"; *NP*, 155), but, at other times, Hudson's examples of "strange and extranatural things in nature" (*ID*, 150) appeal to English folkloric traditions and the Victorian fascination with fairies and freaks.[17] In *Idle Days in Patagonia*, for instance, Hudson claims to have had "personal knowledge" of "mushrooms growing in rings, and the shrinking of the sensitive plant when touched, and will-o'-the-wisps, and crowing hens" (*ID*, 150–51). Whilst all of these phenomena can be explained by modern science—indeed, the touch-sensitive *Mimosa pudica* was described by Carl Linnaeus as early as the mid-eighteenth century—for Hudson they embody "the supernatural in all natural things."[18] His inclusion of a "crowing hen," for instance, reflects a Victorian interest in hermaphroditism, and the reference to mushrooms growing in rings, caused when a central fungus produces small threads (mycelium) in a circular shape, alludes to the English folk belief about fairies dancing in woodland circles.[19]

Hudson's belief in a spiritual, "enchanted" (*ID*, 111) dimension to all life—human and nonhuman—echoes Thoreau's reflections on the limitations of science: "With all your science can you tell how it is, and whence it is, that light comes into the soul?"[20] A paradigmatic example of this struggle between science and mystery in Hudson's early naturalist writings is the author's account of the "strange instinct" (*NP*, 312) of the guanaco of southern Patagonia which, for hundreds of years, had sought out a "dying-place" along a specific portion of the Santa Cruz and Gallegos rivers. In his determination to discover the root cause of this instinct—which he eventually attributes to the animal's residual attraction to areas of dense vegetation in response to ancestral memories of extreme cold and famine—Hudson seems, on the surface, to show more scientific resolve than even Darwin, whom he quotes as admitting to "not at all understand[ing] the reason of this" (*NP*, 313). Nevertheless, it is important to note the weakness of Hudson's apparently "scientific" argument in the chapter he dedicates to the phenomenon in *The Naturalist in La Plata*, which epitomizes the approach of his 1892 book as "chiefly occupied with matters of personal knowledge, seasoned with a little speculation" (*NP*, 233). In the penultimate paragraph of the chapter, Hudson traces the instinct back to a vague "hypothetical origin"—a "habit acquired by the animal in *some* past period of seeking refuge from *some* kind of pain and danger," (*NP*, 322–23; my italics)—before he goes on to "speculate a little further," asking his reader to "imagine" a series of increasingly unsubstantiated suppositions, narrated largely in the conditional tense and punctuated with "if" clauses.

Such an argument supports Fletcher's view that the "dullest passages in [Hudson's] books are his attempts at original scientific investigation or speculation."[21] Earlier in the chapter, we have, by contrast, Hudson at his narrative best, embracing the Romantic predilection for wilderness in his portrayal of the "dense primaeval thicket," where the guanaco takes itself off to die:

> What a subject for a painter! The grey wilderness of dwarf thorn trees, aged and grotesque and scanty-leaved, nourished for a thousand years on the bones that whiten the stony ground at their roots; the interior lit faintly with the rays of the departing sun, chill and grey, and silent and motionless—the huanacos' Golgotha. . . . And now one more, the latest pilgrim, has come, all his little strength spent in his struggle to penetrate the close thicket; looking old and gaunt and ghostly in the twilight; with long ragged hair; staring into the gloom out of death-dimmed sunken eyes. (*NP*, 312–13)

This vision, suffused with a distinctly Conradian gloom, employs anaphora via the repeated conjunction "and" to draw the reader steadily into the benighted world of the dying *guanaco*. The tone is biblical, with the animal anthropomorphized as a bedraggled pilgrim, and the riverine setting metaphorically elevated to the site of Jesus's crucifixion, Golgotha, the name of which derives from the Hebrew word for "skull."[22] The allusion between the dying place of the guanaco and Calvary certainly borders on the sacrilegious, and Hudson implicates his reader further in this blasphemy by invoking our pity not for the death-throes of the son of God, but a diminutive South American camelid as it stares, in its turn, into the abyss. Although at the end of the chapter, Hudson forwards a drier, more scientific interpretation of the "dying place," it is hard for the reader to shake off the emotive force of these earlier passages, which culminate in Hudson's comparison of the animal's instincts to the "superstitious observance of human beings, who have knowledge of death, and believe in a continued existence after dissolution" (*NP*, 312–13). As with Hudson's commentary on the death rituals of the *hornero* in *Birds of La Plata*, throughout these early works the naturalist opens up metaphysical consideration to the nonhuman world, crediting animals with spiritual capabilities that have only recently come to be explored by scientists and theologians.[23] When reflecting on the limitations of specimen collecting in the introduction to *Birds of La Plata*, Hudson makes a similar point, engaging the full etymological force of the noun "animal" (as Paul Badham reminds us, "The Latin word 'anima' means 'soul'") in his observation that "the body is but the case, the habit, and when the life and soul have gone out of it, what is left is nothing but dust" (*BP*, x).[24] Whilst Hudson's reflections revisit well-worn Thomist ideas about animal immortality, they also raise broader questions about the ethics of human interactions with animals, destabilizing established Western boundaries that separate people from nature and anticipating twenty-first century debates about the more-than-human world.

In his account of the guanaco's "dying place," Hudson's appeal to the supernatural and to aesthetic modes of mediating nature (in this case painting) mark his distrust of science as the only way to comprehend the natural world. One of Hudson's biographers, fellow Argentine Ezequiel Martínez Estrada, notes how, for Hudson, a "feather contains, like the flower, miraculous science and art, beauty and exactitude," drawing out the coalescence of mystery and reason spanning all of the naturalist's work.[25] It is fitting that Martínez Estrada should give the example of a feather, for it is the bird, more than any other animal, which embodies for Hudson the spiritual dimension of nature. His persistent fascination with birds is evident across his published work: after his collaboration in *Argentine Ornithology*, Hudson wrote *Birds in a Village* (1893), *British Birds* (1895), *Birds in London* (1898), *Birds and Man* (1901), *Adventures Among Birds* (1913), and *Birds in Town and Village* (1919). In his early South American writings birds are endlessly mysterious: Hudson describes bird migration as a "mysterious thought-baffling faculty, so unlike all other phenomena in its manifestations as to give it among all natural things something of the supernatural" (*ID*, 5). He writes of the "mysterious suspirations" (*NP*, 28) of the rhea, and "the mystery" behind the fiery "dragon eyes" (*ID*, 177) of the Magellanic owl. However, for Hudson, the most esoteric faculty of the bird is its song. Typically dismissing the scientific "method of *spelling* bird notes and sounds" as a "fancy and a delusion" (*ID*, 137–38), Hudson instead attempts to convey bird song through metaphor and kinaesthesia, merging the senses of hearing with those of touch and sight, as with his description of the "soft silvery sounds" (103) of the Yellow-breasted Marsh-bird in *Birds of La Plata*. In the same book he notes that the "delicate melody" of the Chingolo has

a "thin ethereal character, the multitudinous notes not mingling but float-
ing away, as it were, detached and scattered, mere gossamer webs of sound
that very faintly impress the sense" (*BP*, 46). Here Hudson endows sound
with a physicality, albeit one characterized by slightness bordering on the
invisible. The past participles "detached" and "scattered" emphasize the difficulty of categorizing or containing sound, a feature which generates some
fantastical similes, like Hudson's likening of the call of a field finch to the
"finest threads of sound and faintest tinklings, as from a cithern touched
by fairy fingers" (*NP*, 271). The comparison of bird-song to the strains of
an obsolete stringed instrument played by a supernatural being does little to concretize its call for the reader. Earlier in *The Naturalist in La Plata*,
Hudson compares the "thin familiar sounds" of gnats to "horns of elf-land
faintly blowing" (*NP*, 134).[26] As noted earlier, fairy lore was a touchstone
in Victorian England, often invoked "in opposition, or even resistance, to
the progress of science."[27] For Hudson, the semantic field of fairies and the
paranormal afforded one more way to preserve a mythical, enchanted view
of nature—"a nature at once natural and supernatural"—against a deaden-
ing scientific rationalism.[28]

Although one can learn a lot about South American plants and animals
from Hudson's early writings, the aim of these quasi-naturalist works is
not, then, to dispel the mysteries of nature. Instead, like a consummate Romantic, Hudson schools his reader to leave their books at home and experience these mysteries for themselves. Hudson called the "mysterious feelings" (*ID*, 110) arising from close contact with nature "animism," a term he
first uses in *Idle Days in Patagonia* and which remains central to his life-long
study of the complex interactive relationships between humans and nonhumans. Animism, for Hudson, is where mystery and ecology converge; it is
through animism that humans are able to experience, if not fully comprehend, their consonance with the more-than-human world and apprehend
their ethical and ecological obligations toward it.

"In and One with It": Animism, Inter-Relationality, and Posthuman Ecology

In *Idle Days in Patagonia*, Hudson discusses a "long dissertation . . . on whiteness in nature, and its effect on the mind" (104) included in Herman Melville's *Moby Dick*. Rejecting Melville's reading of the phenomenon, Hudson

argues that the "mysterious illusive *something* affecting us in the thought of whiteness" (105) springs "from the animism that exists in us, and our animistic way of regarding all exceptional phenomena":

> Animism here means not a doctrine of souls that survive the bodies and objects they inhabit, but the mind's projection of itself into nature, its attribution of its own sentient life and intelligence to all things—that primitive universal faculty on which the animistic philosophy of the savage is founded. When our philosophers tell us that this faculty is obsolete in us, that it is effectually killed by rationation, or that it only survives for a period in our children, I believe they are wrong, a fact which they could find out for themselves if, leaving their books and theories, they would take a solitary walk on a moonlit night in the "Woods of Westermain," or any woods, since all are enchanted. (*ID*, 110–11)

Once again, we witness Hudson railing against scientific rationalism—all those "books and theories"—and championing alternative modes of knowing nature: the reading of literature (by way of the reference to George Meredith's poem about an enchanted wood), solitary walks, and a "return to an instinctive or primitive state of mind" (*ID*, 205).

At its simplest, Hudson's animism is a form of anthropomorphism ("the mind's projection of itself into nature"), a favored trope of the Romantics. Fletcher notes "a naive, sometimes playful, yet unshakeable belief" in the sentience of the natural world for Hudson.[29] This applies particularly to animals and, in such cases, as Miller notes, Hudson is not above anthropocentrism, often including "whimsical and sentimental anthropomorphic passages where he imagines animals to be carrying on in a distinctly human way."[30] Examples of this can be found in *Birds of La Plata*, for instance, where Hudson cites a description of the "wild human laugh" (*BP*, 240) of the Grey Eagle. Hudson himself acknowledges his propensity to anthropomorphize animals in a chapter of *Idle Days in Patagonia*, dedicated to the story of a "rather small . . . not too fat" (*ID*, 56), curly-haired retriever, Major, noting: "I can even laugh at myself for having allowed an ineradicable anthropomorphism to carry me so far" (*ID*, 67). Hudson encountered Major when staying at an English *estancia* in Patagonia. Despite the dog's ignominious reputation for having killed sheep as a puppy, he becomes a loyal hunting companion of the writer, but falls from grace a second time after he mutilates geese that Hudson had shot for breakfast. Nevertheless, as

Andermann notes, Hudson's psychological account of the dog in *Idle Days in Patagonia* is more complex than that of any human protagonist in the book.[31] The psychodrama of Major's earlier "crime" (*ID*, 62), in which the dog found "the ancient wild-dog instinct . . . hot in his heart" (*ID*, 62); the glory of an early hunting expedition for flamingos during which, against all the odds, he secures a "splendid specimen" for Hudson (*ID*, 61); and the final fall, after which Hudson "looked on him as a poor degraded creature" (*ID*, 67): all these exemplify how Hudson's writing about animals transcends the potential limitations of the figure of anthropomorphism, which has been criticized for its tendency to hold up the human as a "single assertion or essence which, as such, excludes all others."[32] Throughout his writing, Hudson's attention to animal life stories, as in his account of Major or the celebrated "Biography of the Vizcacha" in *The Naturalist in La Plata*, consistently blurs Western demarcations between humans and nonhumans and takes seriously the idea of animal "personhood." His anthropomorphism of animals also goes well beyond the sentimental. Despite the moving tribute to Major, and the powerful role that domestic animals played in the Victorian imagination more broadly, Hudson "hated pets," as Wilson notes, particularly dogs.[33] Rather Hudson recognized that animals have rights (he was the first chairperson of the Society for the Protection of Birds and an anti-vivisectionist), and believed that they exhibit behavior and emotions that need to be acknowledged and respected.[34]

Dominik Ohrem has recently explored issues of animacy in writing about animals. He argues that a key feature of "animating/storying creaturely life is not only about animating the lives of individual beings or species but also, and perhaps even more importantly, about animating *relationality* as such."[35] Far from being a case of simple literary anthropomorphism, Hudson's animistic understanding of nature proposes a profound state of inter-relationality between the human and the nonhuman. In his extended account of animism in the autobiography of his Argentine childhood, *Far Away and Long Ago*, Hudson speaks of the "sense of mystery" inspired by large trees near his home, and of how moonlight shining on the foliage of the acacia, in particular, "made this tree seem more intensely alive than others, more conscious of my presence and watchful of me."[36] Such a statement brings together Hudson's self-styled "primitivism"—his engagement with pre-modern or non-Western thinking about the sentience of nature—and his prescient ecological imagination, which considers plants and animals as sapient beings, intimately related to humans. It also embeds Hudson in a specifically Amerindian worldview, which has variously been termed "per-spectivist" or "multinatural," and which, as Eduardo Viveiros de Castro notes, is predicated on the "virtually universal Amerindian notion . . . of an original state of undifferentiation between humans and animals": "The differentiation between 'culture' and 'nature' . . . is not a process of differentiating the human from the animal, as in our own evolutionist mythology. The original common condition of both humans and animals is not animality but rather humanity."[37]

This definition strikes a chord with the view of nonhuman nature expressed in much of Hudson's early South American nature writing. Hudson observes that what he terms "animism" arises not when someone is "outside and above the natural phenomenon, but in and one with it" (*ID*, 112), suggesting a collapsing of ontological distinctions between people and nature, whether rocks, or trees, or animals. One of the most important instances of animism across all of Hudson's work is in "The Plains of Patagonia," the penultimate chapter of *Idle Days in Patagonia*. The chapter is insistently marked by a blurring of boundaries, with notable shifts from not only human to animal, but vegetal to mineral, as in the metaphor figuring the motionless vegetation of the plains as "unmoving as if carved out of stone" (*ID*, 199). Hudson speaks repeatedly of the undiminished greyness of Patagonia:

> Everywhere through the light, grey mould, grey as ashes and formed by the ashes of myriads of generations of dead trees, where the wind had blown on it, or the rain had washed it away, the underlying yellow sand appeared, and the old ocean-polished pebbles, dull red, and grey, . . . the hills . . . were clothed in the grey everlasting thorny vegetation. How grey it all was! hardly less so near at hand than on the haze-wrapped horizon, where the hills were dim and the outline blurred by distance. (196–97)

Grey is an intermediary color between black and white, and its dominance in this passage, alongside references to "haze," dimness and blurring, present Patagonia not as a place of certainty but of states-of-becoming, where trees turn to mould, pebbles to sand, and where Hudson too undergoes a transformation. In "The Plains of Patagonia," Hudson relates how he fell into the habit of taking rambles across the desolate landscape, and "finding and using [a small grove] as a resting-place every day at noon" (*ID*, 197). During these excursions, Hudson's mind is "suddenly transformed" into a "state [...] of *suspense* and *watchfulness*" (*ID*, 199) consonant, in his view, with the animistic outlook of the "pure savage" who "is in perfect harmony with nature" (*ID*, 205). As noted in the introduction, Castro-Klarén regards Hudson's formulation of animism as having clear parallels with the shamanic trance state.[38] Like the shaman, Hudson undergoes a form of metamorphosis in Patagonia. He speaks of the transformative state of animism, observing that his alteration in the solitudes of Patagonia was "as great and wonderful as if I had changed my identity for that of another man or animal" (*ID*, 199). Although here and earlier, Hudson holds back from entirely relinquishing his human subjectivity, his metamorphosis is consistently figured in this chapter as cross-species. He notes his "animal-like" (*ID*, 198) return to the favored grove of trees, notably also a haunt of a herd of wild animals, and describes his "revived instinct" as "purely animal in character" (*ID*, 200). A central image of metamorphosis in the chapter is Hudson's adoption of the metaphor of a caterpillar encased in a cocoon to describe the process of estrangement from nature in humans. Only "miraculous moments" (*ID*, 204)—those flashes of animism—allow the cocoon to dissolve, and a rapprochement with the more-than-human world to occur.

Hudson's early South American writings are marked by these "miraculous" encounters with the nonhuman other. In the second chapter of *Idle Days in Patagonia*, Hudson recalls how, the night following his calamitous leg injury, he unknowingly shared his bed with a venomous snake, only discovering it the next morning: "My hospitality had been unconscious, nor, until that moment, had I known that something had touched me, and that virtue had gone out from me" (*ID*, 25). Christoph Irmscher has noted a morbid fascination with snakes across eighteenth- and nineteenth-century natural history.[39] Hudson shared this interest, as Miller notes, particularly in snake mythology, but the serpentine bedfellow of *Idle Days in Patagonia* lacks any sinister intent.[40] Instead, this unexpected intimacy affords Hudson the chance to be "in and one with" (*ID*, 112) the nonhuman world, just as a later snake encounter precipitates a profound realization in Hudson about his place in the web of life and of "kinship with it in all its appearances, in all organic shapes, however different from the human."[41] Hudson's "animism" can be understood as a kind of posthuman ecology, where humans and nonhumans are not regarded as separate but as relational and co-constitutive. For the Argentine author, animism takes place when the threshold between the human and the nonhuman is crossed: when we are penetrated by "a sense of the *thing itself*—of the tree or wood, the rock, river, sea, mountain, the

soil, clay or gravel, or sand or chalk, the cloud, the rain, . . . as if the quality of the thing itself had entered into us, changing us, affecting body and
mind"; or, conversely, when we enter into the consciousness of the nonhuman subject, the acacia tree in *Far Away and Long Ago*, or the "quaint furry"
creature imagined by Hudson in *The Naturalist in La Plata*, which "take[s]
in the unfamiliar flavour of a human presence from the air" (*NP*, 362–63,
italics in original) before scampering away.[42]

Timothy Morton has defined the "ecological thought" as "a practice and
a process of becoming fully aware of how human beings are connected with
other beings— animal, vegetable, or mineral."[43] Hudson's animist imaginary was, in the late nineteenth century, at the vanguard of such inter-
relational thinking. Whilst retaining a Romantic suspicion of science, his
early South American writings also reflect a strong commitment to ecology—the understanding of nature "as a unified whole made up of complex interrelationships," which, as Andrea Wulf argues, can be traced at
least as far back as Humboldt, but also resonates with contemporary ecological thinking.[44] In *Birds of La Plata* Hudson acknowledged the "inter-
lacing relations" between "the lives of all living creatures" (*BP*, 252), and
voiced concerns over habitat loss and extinction in terms that are painfully
familiar to the twenty-first century reader. He believed in a "commensal-
ism on earth from which the meanest organism is not excluded."[45] Arocena argues that Hudson "founded an extremely original line of ecological
thought which is still completely valid in our world today."[46] Hudson's animism also remains valid, and continues to have potentially radical consequences for our engagements with nature, raising as-yet unanswered philosophical, ethical, and ecological questions about how humans treat the
more-than-human world.

Conclusion

Hudson sums up his *Idle Days in Patagonia* as a "record of what I did not do"
(17). Following on from this characterization, Hudson's early South American nature writing might be described as a *record of what he did not know*:
a nature that he accepted, in familiar Romantic terms, as mysterious and
supernatural. Yet this "not knowing" also signals the prescience of Hudson's
ecological thinking, which anticipates what the environmentalist Wes Jackson has more recently characterized as an "ignorance-based worldview":

The modern scientific program has held that we must act on the basis of knowledge, which, because its effects are so manifestly large, we have assumed to be ample. But if we are up against mystery, then knowledge is relatively small, and the ancient program is the right one: Act on the basis of ignorance. Acting on the basis of ignorance, paradoxically, requires one to know things, remember things—for instance, that failure is possible, that error is possible.[47]

It is precisely Hudson's openness to mystery in the natural world, and to not knowing and not thinking—the animistic "state . . . of *suspense* and *watchfulness*" (*ID*, 199) that he embraces in the plains of Patagonia—that leads to some of his most important realizations about nature. In *Birds and Man* (1901) he observes that "if the mystery of life daily deepens, it is because we view it more closely and with clearer vision."[48] For Hudson, mystery arises not from lack of knowledge about nature but a recognition that this knowledge is more complex and diffuse than we can fully comprehend. Hudson's earliest works on South America are also complex and diffuse, melding natural science, mythology, and gaucho imaginaries, and marked by "animism," speculation, and intuition. In *Idle Days in Patagonia*, Hudson describes how he would watch flies from his sick bed, "flitting, sylph-like things," whose "whirling" flight resembled "strange characters in the air, all forming a strange sentence—the secret of secrets!" (*ID*, 18). Hudson's naturalist writings sometimes feel like the "mazy dance" (*ID*, 18) of these flies, drawing us in with the promise of illumination before "mocking [our] power to grasp them, and darting off again at a tangent" (*ID*, 18). And yet the "secret of secrets" (*ID*, 18) is revealed: nature may be intrinsically mysterious, but we must continue to observe it, to imagine it, to be "in and one with it" (*ID*, 112).

NOTES

1. As recalled by Ford Madox Ford, cited in Miller, *W. H. Hudson and the Elusive Paradise*, 2.
2. Hudson, *Naturalist in La Plata*, 6. Page numbers from this book will hereafter be included in the text in parentheses following the abbreviated title, e.g., (*NP*, 6). See Wilson, *Living in the Sound of the Wind*, 227.
3. Cited in Wilson, *Living in the Sound of the Wind*, 227.
4. Gómez, "Introducción," 20, 11; Wilson, "Colonial's Revenge," speaks of Hudson's "in-between position," 5.
5. Arocena, *William Henry Hudson*, 2. Canby speaking of Hudson, cited in Gómez, "Introducción," 11.

6. Hudson, *Idle Days*, 110–11. Page numbers from this book (*ID*) will hereafter be included in the text in parentheses.

7. See Castro-Klarén, "Recorridos chamánicos."

8. Wilson, *Colonial's Revenge*, 20.

9. See Tomalin, *W. H. Hudson*, 90–91 and 237–40; and Jameson, *Finding W. H. Hudson*, 8–12.

10. See Wilson, *Living in the Sound of the Wind*, ch. 11.

11. Arocena, *William Henry Hudson*, 4.

12. Hudson, *Birds of La Plata*, ix. Citations from this text (*BP*) will hereafter be given parenthetically. See also Jameson, *Finding W. H. Hudson*, 7–8.

13. Andermann, "Pulsión animal," 115.

14. For an account of this movement, see Blanco and Page, "Introduction," 271.

15. For instance, Rousseau, *Discourses*, 124: "the more new knowledge we accumulate, the more we deprive ourselves of the means of acquiring the most important knowledge of all."

16. For a comparison with the Romantics, see Fletcher, "The Creator of Rima"; Walker, "W. H. Hudson," focuses on the parallels between Hudson and the New England tradition.

17. As a naturalist, Hudson was not alone in his interest in non-normative nature. Charles Waterton, well-known to Hudson, notably kept a menagerie of abnormal animals in his country estate, including a crowing hen and a sheep with a horn growing from its ear; see Edginton, *Charles Waterton*, 189.

18. Hudson, *Book of a Naturalist*, 211. See Miller, *W. H. Hudson*, 43, for a discussion of this important phrase that lends itself to the title of the second chapter of Miller's book.

19. For a discussion of hermaphroditism in Victorian culture see, Tromp and Valerius, "Introduction," 3. See Silver, *Strange and Secret Peoples*, 36, for a reference to Victorian beliefs about fairy rings.

20. Thoreau, *Journal Volume II*, 307.

21. Fletcher, "The Creator of Rima," 34.

22. *OED Online*, s.v. "Golgotha," accessed May 2, 2021, https://www-oed-com.

23. See, for instance, the account of religious feeling among primates in Goodall, "Primate Spirituality," 1,303–04; Linzey and Yamamoto, *Animals on the Agenda*.

24. Badham, "Do Animals have Immortal Souls?," 181.

25. Martínez Estrada, "Estética y filosofía de Hudson," 42.

26. This line is from Tennyson's "The Splendour Falls."

27. Brown, *Fairies*, 103.

28. Hudson, *Far Away and Long Ago*, 308.

29. Fletcher, "The Creator of Rima," 29.

30. Miller, *W. H. Hudson*, 41.

31. Andermann, "Pulsión animal," 112.

32. de Man, *Rhetoric of Romanticism*, 241.

33. There is a vast bibliography on pets in Victorian Britain. See, for instance, Howell, *At Home and Astray*; Wilson, *Living in the Sound of the Wind*, 94–95 (94).

34. Wilson discusses his involvement with the anti-vivisection movement in *Living in the Sound of the Wind*, 215; The best account of his relationship with the Society for the Protection of Birds is in Jameson, *Finding W. H. Hudson*.

35. Ohrem, "Animating Creaturely Life," 10.

36. Hudson, *Far Away and Long Ago*, 242–43.

37. Viveiros de Castro, "Cosmological Deixis," 471–72.

38. Castro-Klarén, "Recorridos chamánicos," 31.
39. Irmscher, *Poetics of Natural History*, ch. 4.
40. Miller, *W. H. Hudson*, 67.
41. Hudson, *Book of a Naturalist*, 23.
42. Hudson, *A Hind in Richmond Park*, 33–34
43. Morton, *Ecological Thought*, 7.
44. Wulf, *Invention of Nature*, 307
45. Hudson, *Birds and Man*, 245.
46. Arocena, *William Henry Hudson*, 8.
47. Jackson, "Toward an Ignorance-Based Worldview," 22.
48. Hudson, *Birds and Man*, 245.

BIBLIOGRAPHY

Andermann, Jens. "Pulsión animal: Zooliteratura y transculturación en W. H. Hudson." In *Entre Borges y Conrad: Estética y territorio en William Henry Hudson*, edited by Leila Gómez and Sara Castro-Klarén, 107–25. Iberoamericana; Vervuert, 2012.

Arocena, Felipe. *William Henry Hudson. Life, Literature and Science*. Translated by Richard Manning. McFarland, 2003.

Badham, Paul. "Do Animals Have Immortal Souls?" In *Animals on the Agenda: Questions about Animals for Theology and Ethics*, edited by Andrew Linzey and Dorothy Yamamoto, 181–89. University of Illinois Press, 1998.

Blanco, María del Pilar, and Joanna Page. "Introduction to Section V." In *Geopolitics, Culture, and the Scientific Imaginary in Latin America*, edited by María del Pilar Blanco and Joanna Page, 271–74. University of Florida Press, 2020.

Brown, Nicola. *Fairies in Nineteenth-Century Art and Literature*. Cambridge University Press, 2001.

Castro-Klarén, Sara. "Recorridos chamánicos: Sobre el afecto cognitivo en Arguedas, W. H. Hudson y Deleuze y Guattari." *Revista de crítica literaria Latinoamericana* 75 (2012): 27–50.

de Man, Paul. *The Rhetoric of Romanticism*. Columbia University Press, 1984.

Edginton, Brian W. *Charles Waterton: A Biography*. Lutterworth Press, 1996.

Fletcher, James V. "The Creator of Rima: W. H. Hudson: A Belated Romantic." *Sewanee Review* 41 no. 1 (1933): 24–40.

Gómez, Leila. "Introducción." In *Entre Borges y Conrad: Estética y territorio en William Henry Hudson*, edited by Leila Gómez and Sara Castro-Klarén, 7–30. Iberoamericana; Vervuert, 2012.

Goodall, Jane. "Primate Spirituality." In *Encyclopedia of Religion and Nature*, edited by Bron Taylor, 1,303–06. Continuum, 2005.

Howell, Philip. *At Home and Astray: The Domestic Dog in Victorian Britain*. University of Virginia Press, 2015.

Hudson, W. H. *A Hind in Richmond Park*. Dent, 1923.

Hudson, W. H. *Birds and Man*. Dent, 1923.

Hudson, W. H. *Birds of La Plata*. Dent, 1923.

Hudson, W. H. *The Book of a Naturalist*. Dent, 1923.

Hudson, W. H. *Far Away and Long Ago: A History of My Early Life*. Dent, 1923.

Hudson, W. H. *Idle Days in Patagonia*. Dent, 1923.

Hudson, W. H. *The Naturalist in La Plata*. Dent, 1923.

Irmscher, Christoph. *The Poetics of Natural History: From John Bartram to William James*. Rutgers University Press, 1999.

Jackson, Wes. "Toward an Ignorance-Based Worldview." In *The Virtues of Ignorance. Complexity, Sustainability, and the Limits of Knowledge*, edited by Bill Vitek and Wes Jackson, 21–36. University Press of Kentucky, 2008.

Linzey, Andrew, and Dorothy Yamamoto, eds. *Animals on the Agenda: Questions about Animals for Theology and Ethics*. University of Illinois Press, 1998.

Martínez Estrada, Ezequiel. "Estética y filosofía de Hudson." In *Antología de Guillermo Enrique Hudson con estudios críticos sobre su vida y obra*, 33–46. Losada, 1941.

Miller, David. *W. H. Hudson and the Elusive Paradise*. Palgrave Macmillan, 1990.

Morton, Timothy. *The Ecological Thought*. Harvard University Press, 2010.

Ohrem, Dominik. "Animating Creaturely Life." In *Beyond the Human-Animal Divide: Creaturely Lives in Literature and Culture*, edited by Dominik Ohrem and Roman Bartosch, 3–19. Palgrave Macmillan, 2017.

Rousseau, Jean-Jacques. *The Discourses and Other Early Political Writings*. Edited and translated by Victor Gourevitch. Cambridge University Press, 2003.

Silver, Carole G. *Strange and Secret Peoples: Fairies and Victorian Consciousness*. Oxford University Press, 1999.

Thoreau, David Henry. *Journal: Volume II*. Edited by Bradford Torrey. Houghton, Mifflin, 1906.

Tomalin, Ruth. *W. H. Hudson: A Biography*. Faber and Faber, 1982.

Tromp, Marlene, with Karyn Valerius. "Introduction: Situating the Victorian Freak." In *Victorian Freaks: The Social Context of Freakery in Britain*, edited by Marlene Tromp, 1–18. Ohio State University Press, 2008.

Viveiros de Castro, Eduardo. "Cosmological Deixis and Amerindian Perspectivism." *The Journal of the Royal Anthropological Institute* 4, no. 3 (1998): 469–88.

Walker, John. "W. H. Hudson, Argentina, and the New England Tradition." *Hispania* 69, no. 1 (1986): 34–39.

Wilson, Jason. *Living in the Sound of the Wind: A Personal Quest for W. H. Hudson. Naturalist and Writer from the River Plate*. Constable, 2015.

Wilson, Jason. *W. H. Hudson: The Colonial's Revenge*. Working Papers 5. Institute of Latin American Studies, 1981.

Wulf, Andrea. *The Invention of Nature: The Adventures of Alexander von Humboldt, the Lost Hero of Science*. John Murray, 2016.

CHAPTER 8

Estanislao Severos Zeballos, or, Nineteenth-Century Argentina's Environmental Unconscious

AARTI S. MADAN

> El cultivo de los árboles conviene a un país pastoril
> como el nuestro, porque no solo la arboricultura se une
> perfectamente a la ganadería, sino que debe considerarse
> un complemento indispensable. . . . La Pampa es
> como nuestra República, tala rasa. Es la tela en la que ha
> de bordarse una nación. Es necesario escribir sobre ella.
> ¡Árboles! ¡Planten árboles!
>
> The cultivation of trees benefits a pastoral country like
> ours, because not only does arboriculture combine
> perfectly with cattle farming, but rather it ought to be
> considered an indispensable complement. . . . The Pampa
> is like our Republic, clear cut. It is the fabric on which
> we must stitch a nation. It is necessary to write upon her.
> Trees! Plant trees!
>
> —DOMINGO FAUSTINO SARMIENTO (1870)

On August 29, 1900—some thirty years after sitting Argentine president Domingo Faustino Sarmiento urged his compatriots to punctuate the landscape with trees—Argentine statesman, lawyer, geographer, and quintessential *letrado* Estanislao Severos Zeballos realized an environmental dream he had long imagined: at his behest, the Consejo Nacional de Educación approved Argentina's Día del Árbol, which has been celebrated annually ever since.[1] By institutionalizing Sarmiento's call-to-arms, Zeballos left behind a deep-rooted arboreal legacy consisting of not only a century's worth of trees but also annual environmental education campaigns across the nation. Yet, as part of Argentina's Generación del Ochenta, Zeballos sought progress in positivist thought and military machinations that would wrangle the space of nature into the order of gridlines. His 1878 *La conquista de quince mil leguas* amounts to the ideological construction (Viñas) and visual conquest (Andermann) of the desert, while the genocidal campaign it orchestrated against Argentina's Indigenous communities—officially called the Conquista del Desierto (1879–85)—has been described as a war against nature (Hudson).[2] Only ruins remained after the desert campaign, Sarmiento's *tala rasa* converted into Zeballos's *tabula rasa*.

How do we make sense of wholly contradictory nation-building projects? Drawing on Lawrence Buell's understanding of the *environmental unconscious*, this chapter unpacks nineteenth-century Argentina's paradoxical approach to production and preservation of nature. Buell traces the "the history of interwoven controversy and advocacy of civilizationist and naturist persuasions" arising from the US's rapid environmental transformation over some two hundred years. I believe that his historicizing impulse equally serves the Latin American context; he hints as much by suggesting that such latticework is "less unique than symptomatic of the modernization process" and merits study "beyond any one national instance."[3] By attending to forms of environmental awareness that privilege both anthropocentric and ecocentric ethics—an ostensibly irresoluble conflict that structures the environmental unconscious—Buell proposes a reading practice that resists "binary naïveté."[4]

Rather than a labyrinth of antagonisms, this path allows us to draw out both positive and negative manifestations of the environmental unconscious. As I will show, in the former, humans become attuned to the physical environment and to linkages with the more-than-human-world; the latter evokes the impossibility of connection at any level, be it perception or expression. Buell's theory can thus be applied to Zeballos's seemingly disparate environmental values to illuminate the unevenness of postcolonial ecologies in two primary ways. First, Zeballos's work lends itself to a kind of nuanced ecocritical historicism that strives to raise rather than resolve questions. Second, and relatedly, his writings shed light on the ways in which nineteenth-century Argentina's environmental unconscious emerges circuitously and politically through aesthetic practices.

Scholarship on nineteenth-century Latin America has largely ignored Zeballos's contributions to ecological thinking or, for that matter, to anything beyond geography in service of genocide. While a 1916 political satire in *La Nación* aligned Zeballos's political death with a defunct literary life, more recently Daniel Balderston described him as "a sort of Buffalo Bill of the Argentine south," "a vainglorious braggart and unscrupulous collector of Indian curiosities."[5] Failure and controversy checkered Zeballos's career, whether as a lawyer, professor, geographer, or—perhaps most conspicuously—as a literary writer and career politician.

Yet Zeballos's footprint was anything but insignificant. He brokered border disputes with Argentina's neighboring nation-states and facilitated mediation by the United States; he traveled extensively both nationally and internationally into spaces mapped and unmapped; he founded institutions like the Sociedad Científica Argentina (1872)—at just sixteen years old—and later the Instituto Geográfico Argentino (1879), all the while founding and editing publishing organs on both sides of the Atlantic, including the *Anales de la Sociedad Científica Argentina*, the *Boletín del Instituto Geográfico Argentino*, and the *Revista de Derecho, Historia y Letras*. In many ways, Zeballos's intellectual formation, elite networks, and territorial knowledge matched and even informed those of better-known contemporaries like Domingo Faustino Sarmiento and Francisco Moreno.

For instance, Zeballos published *La conquista de quince mil leguas* at the request of Minister Julio Argentino Roca, who sought to convince members of Congress to finance the Campaña del Desierto. As highlighted by Fermín Rodríguez, this "panfleto ideológico, manual geográfico y apunte histórico" (ideological pamphlet, geographic manual and historical note) was written by the twenty-four-year-old Zeballos in less time that the three-month-long war he choreographed.[6] Yet it ultimately received accolades, critical acclaim, and positive reviews from Sarmiento himself, who suggested that Zeballos's inaugural book ought to accompany soldiers to the warzone.[7] Similarly, Zeballos's staunch support for Moreno's expedition to Patagonia, which he describes in detail in *Conquista*, launched Moreno's storied career as an explorer and pseudo-conservationist who founded Argentina's first national park.

By the early 1900s, Zeballos's orbit of hemispheric statesmen invested in nature and nation—for myriad and often muddied motives—included former US president Theodore Roosevelt and Cuban independence fighter José Martí. The transnational reach of Zeballos's political, literary, and scientific network, the longevity and variety of his nation-building activities, and the capaciousness of his environmental writings invite us to do more than critique his obvious failures, but rather to probe his contributions to evolving understandings of nature vis-à-vis aesthetic and economic practices. By examining Zeballos's role in replacing the term "nature" with the more utilitarian "natural resources" alongside his foresight about erosion, flooding, and aridity—and simultaneously making sense of his role in literal genocide and literary eternity for Argentina's Indigenous inhabitants—I aim to interrogate his contradictory conservationist leanings: did he understand the crime of his militarism, or is his ambiguity a kind of spectral presence that holds his own actions at bay, unconsciously calling them into question? Do his aesthetics obscure genocide while safeguarding the environment *and* the Argentina he wants to build in industry yet restore in narrative?[8]

To broach these questions, I will explore the ways in which Zeballos's environmental unconsciousness manifests in his writings from approximately 1878 to 1886, in particular the extermination blueprint *La conquista de quince mil leguas* (1878), the geographies *Viaje al país de los araucanos* (1881) and *La rejion del trigo* (1883), and the Indianist novels *Callvucurá y la dinastía de los Piedra* (1884) and *Painé y la dinastia de los zorros* (1886). My focus on this period is purposeful. In addition to encompassing his most prolific years, it marks a moment of continent-wide expansion to foment an agrarian frontier on which large tracts of land—typically in the interior of the new republics— were deforested and reconfigured with new forms of agriculture.

Zeballos's discursive practice played a key role in each stage of this neo-colonial conquest, which Fernando Navarro describes as a southern Manifest Destiny.[9] Yet, I will demonstrate that a close reading of Zeballos's literary register reveals an incipient anti-imperial ecology that nourishes his political project. Though Eduardo Pous Peña designates Zeballos as one of Argentina's first conservationists, his ecological historicization greenwashes Zeballos's legacy without problematizing his complicity in the dark history of cultural genocide and environmental degradation.[10] And while Pablo Ernesto Suárez inversely confirms Zeballos's protagonism in Latin American extractivism, I believe we can complicate such readings by examining his prescient observations on causal links between anthropogenic alterations of ecosystems and "natural" disasters.[11] Zeballos's contradictions, I will show, come to light through his aesthetics, which reveal an arrhythmic evolution in his environmental unconscious that is characteristic of other hemispheric ecological efforts.

Archetypical Zeballos: Positivist Precision

The very title of *La conquista de quince mil leguas* portends telluric transformation, a conquest not only of leagues of space but also of legions of subjects, wherein the displacement and disappearance of Indigenous communities allows land to be transformed into property. Zeballos evinces the capitalist discourse of enclosure in *Conquista's* early pages, citing Article 8 of the legal treatise that authorized Roca's military expedition. The law's ratification inaugurates Argentina's discursive and material entry into modernity as the state grants subjecthood to the commanders, officials, and soldiers that conquer the desert, individuals who have little attachment to the land they will soon be deeded as their own. At the same time, the law denies that very subjecthood to the nomadic tribes whom it dehumanizes by way of dispossession.[12] If preenclosure and precapitalist "inhabitancy names a constitution of the human as coexisting with a community and within an environment," then Zeballos participates in a contradictory modernization project in which environmental awareness and place-based connectivity decrease while territorial knowledge and space-based legibility increase.[13]

This geographic legibility requires the kind of qualitative and quantitative precision that Zeballos applauds at every turn, for only taxonomic exactitude affords his work—and by extension the state—an air of truth and the possibility of sedentarization. "Hyperfocus," Buell notes, "is, in part, a recourse for bringing a semblance of order to . . . a region whose spatial arrangements, geographical border, and social organization" can only be described sketchily.[14] This positivist focus on order and progress unfolds in *Conquista* as citations, charts, and tables that enumerate river volumes and other statistics with what Buell might describe as "minute and atomized set-piece descriptions."[15] With rudimentary prose, Zeballos cites those travelers he deems most precise, advancing through a century of bibliography in twenty pages. Amid roughly twenty mentions of praise for precision or exactitude—or, on the contrary, criticism of a map that is "deplorablemente inexacto" (deplorably inexact)—one citation stands out for its length: a September 15, 1875 petition from a young explorer named Francisco Moreno, indeed, the very "Perito" Moreno who would years later inscribe Patagonia onto the national map.[16]

In the petition—read to the Sociedad Científica Argentina (SCA), included in its minutes, and excerpted by Zeballos into *Conquista*—Moreno requests funds to realize an expedition along the Río Negro, a venture sure to satisfy the SCA's primary objective: to foster the sciences by exploring land theretofore unknown (to Creole eyes).[17] Zeballos introduces the petition's excerpts by situating himself as a catalyst who advised Moreno from the idea's inception and urged him to travel under the auspices of the SCA, which Zeballos had founded three years prior in 1872.[18]

By recounting his fervent support for Moreno's geographic venture, Zeballos centers himself in the science of statecraft and instantiates a calculus that James C. Scott describes in *Seeing Like a State*, wherein the state's efforts at sedentarization—to permanently settle lands, to colonize—equal its attempts to make society legible. This legibility manifests from above, what Scott calls a "synoptic view," indeed a leitmotiv seen in the imperial scene of "monarch-of-all-I-survey" (Pratt), the colonialist trope of the "sweeping visual mastery of a scene" (Spurr), or in the "perspectival vision of European maps" (Taylor).[19] Each of these ocular engagements with the physical environment dovetails with the distance characteristic of Buell's notion of hyperfocus, which conversely lessens the environmental unconscious. To *see* the land does not imply attunement—which requires other senses—but rather "the modernizing need to undo the premodern state's partial blindness."[20] As with the administrative mechanisms developed by the Spanish Crown (Uriarte) and the British Empire (Marzec), *La conquista de quince mil leguas* privileges expeditions and excerpts that look to metrics to imply certainty about the seen environment.[21] The Zeballos that authored *Conquista* is, then, the archetype we know—the paradigmatic fin-de-siecle statesman who sought to overlay territory with statistical reliability and transform land into an entity to be cultivated and *improved*, a key term that rationalized privatization.[22]

Ecoambiguous Zeballos: Creative Cultivation

Though only five years separate Zeballos's *La conquista de quince mil leguas* (1878) and *La rejion del trigo* (1883), they differ dramatically in their narrative approach to time, space, and certainty. In *Conquista*, he incorporates the past into his text by translating colonial-era "data into narrative" while relying on "the pretense of precision" to detail the burgeoning nation-state's strategic hydrography and geography.[23] This work orchestrates a future war. *La rejion del trigo*, on the other hand, presents readers an aesthetically distinct experience. The narration is nonlinear and literary; privatization has segued into productivity. In this second installment of a three-part geography titled *Descripcion amena de la geografía argentina*, Zeballos has matured into a storyteller who uses memories and anecdotes to recall changes in the landscape, while he looks to reflection to speculate on a future of agricultural productivity. As Suárez has shown, the title itself showcases Zeballos's epistemological uncertainty. Zeballos registers the presence and utility of rivers, lakes, and streams to facilitate agricultural exploitation of Santa Fe, which in 1883 is not yet the "region of wheat" that the title portends. The title of the book, Suárez contends, conjures a vision of optimism and productivity grounded in the extractivist projects of the 1880s and departing from the difficulties of 1852–1880, when territorial knowledge functioned more in service of armed expansion and less toward cultivating and exporting agrarian goods.[24]

To Suárez's suggestion of a fluvial future, I would add that Zeballos is Janus-faced in his approach: he reminisces about the past in ways that almost formulaically animate Renato Rosaldo's notion of "imperialist nostalgia," in which colonial agents paradoxically yearn for "the very forms of life they intentionally altered or destroyed."[25] As though to negate his complicity, Zeballos now assumes a vantage point from the ground rather than from the surveillance of above; his representation of both the human and nonhuman world signal an incipient environmental awakening that stands in contrast to the distance and foreshortening in his earlier work.

In a chapter on the *colonia* Candelaria, for instance, Zeballos describes a horrific scene from 1864: the fields have been razed, the cattle captured, the families held captive, and young people sacrificed. Silence prevails in this space of destruction as death consumes the lands.[26] Zeballos equates life with shepherding, yet both living and livelihood are impossible without the guard of the army, which at that moment was fighting on the frontlines of the Guerra del Paraguay.[27] After detailing the heavy stillness of the scene he beckons readers to leave it behind—"Abandonamos, pues, aquellos *pagos de* Arequito" (Let us forget, then, the price of Arequito).[28] Yet he suggests that his beloved town will accompany him his whole life, for Arequito was the "teatro" of his first impressions *santesfesinas*, which included happiness (*sonrientes*) while he pursued a deer or an ostrich on his little racehorse, curiosity (*estrañas*) when he contemplated the bones of giants exhumed by the waters of the deep gorges, and horror (*pavorosas*) upon hearing the shrieks of the Indians vibrating in the air and cleaving his heart like a poisoned dart.[29] Zeballos's relationship to the terrain and its native inhabitants, both nonhuman and human, emerges as dominance and subservience: he chases defenseless animals, a human predator mounted upon an animal that he controls with a whip; yet he stands in awe at the sight of exhumed fossils in the Río Carcarañá, bones that the locals attribute to an extinct race of giants. He fears for his life as his ears presume danger, its sonic traces vibrating with both silence and screams.

Zeballos's affective response to nature in this scene reveals in miniature what can be traced across an entire decade-long corpus in which he depicts space and subject, environment and inhabitant: he is simultaneously attracted to and repelled by the Argentine land's undeveloped state. In this sense, his evocation of the physical environment of Arequito is not necessarily *just* about that one context but rather illustrative of a more complex engagement with environment than initially appears. For Zeballos—and other figures engaging in acts of writing and reading about nature—there appear to be parallel processes of (1) environmental awakening, what Buell calls "retrievals of physical environment from dormancy to salience" and (2) "of distortion, repression, forgetting, inattention."[30] While Buell reminds us that most people "may not even want to attune themselves self-consciously to their environments lest this produce sensory overload, confusion, and despondency," Zeballos actively does the hard work of remembering, processing, and representing even though the environment foregrounds the impossibility of representation.[31] While his narration first empties the land, he does not flee it; he returns to it, he returns sound to space, which takes on new meaning as his environmental unconscious shifts from occlusion to opening.

This rendition of "feeling one's way through dangerous territory" positions Zeballos in a genealogy of hemispheric writers for whom, following Buell, "topophilia and topophobia alternate, clash, and fuse" in ways that neutralize the landscape of fear.[32] Put another way, Zeballos reveals positive attachment to place—and respect for its destabilizing energy and creative force. Only the literary dimension can tease out this tension, which manifests as he fills some two-and-a-half pages with the horrors of the past and then segues to the promise of the future: his awe in the face of nature's sublimity morphs into praise for its instrumentality. Between 1864 and 1878, he tells us, "la transformacion habia sido completa!" (the transformation had been complete!).[33] The space of nature prior to an elliptical section break, which visually marks the changing relationship between humans and environment, appears as a "solitario desierto" (solitary desert), its population limited save for the *araucanos* who traversed the plains.[34] After the ellipses, Zeballos observes that "las hordas de salvajes" (savage hordes) have abandoned the terrain, while immigrants of all nationalities have established roots as colonizers.[35]

Taking a page from Sarmiento's playbook, Zeballos employs Humboldtian geographical discourse to attract those very European immigrants in

Viaje al país de los araucanos (1881) and to educate them about the national land and language.[36] In the early pages of *Viaje*, nature, nation, and narration collide as he promises to replicate "la escuela fundada por Humboldt, que ofrece las lecciones de la Ciencia clareadas por la fosforecencia de una alma ardorosa y de una imaginacion brillante" (the school founded by Humboldt, which offers the lessons of Science illuminated with the phosphorescence of an arduous soul and brilliant imagination).[37] Zeballos admits that he purposefully gives his "especulaciones literarias un tinte científico" (literary speculations a scientific tint) while also founding institutions, publishing studies, and encouraging explorations, all to serve as an example for the nation's youth.[38] That he perceives his travelogue to be literary speculations with a scientific tint—rather than science with a literary bent—reveals his narrative priorities. Highlighting the bidirectional nature of his account, moreover, he writes for his "compatriotas y al estranjero" (compatriots and foreigners) so both can learn about the Pampas.[39] Didactic and marketable, this geography will attract national and international readership by appealing to a literary register that departs from the taxonomic impulse driving *Conquista*, a text consumed less by the general reading public and more by the military apparatus it was conceived by and for. This time, Zeballos aspires to inform with picturesque descriptions that will shift readers' perspective from indifference to curiosity.[40]

At the intersection of politics and poetics, of ecology and cosmology, Zeballos imagines a community while acknowledging that the rarity of his wilderness enhances its value. His nationalism resounds as he dedicates *Viaje* to his "Pátria" (Fatherland) and signals his desire both to be "útil a su Pais" (useful to his Country) and "honrar las letras argentinas" (to honor Argentine letters).[41] This two-pronged effort immediately hooks onto nature with an epigraph attributed to V. H. that poses an "I" as the mysterious lover of nature, someone who speaks with the wind and the trees, someone known by the fields and the jungles:

> Si, yo soy el amante misterioso
> De la Naturaleza; el camarada
> De la amarilla flor que se columpia
> En la vieja pared; yo soy quien habla
> Con el viento y los árboles: conócenme
> Los campos y las selvas . . .

Yes, I am the mysterious lover
Of Nature; the comrade
Of the yellow flower that sways
Along the old wall; I am who speaks
With the wind and the trees: the fields
And the jungles know me . . .

Zeballos's epigraph evinces not only a desire to be attuned to nature but *by* nature, to awaken and to activate an ecological conscience in readers.

Yet his duplicity shines just pages later when *knowing nature* again functions in service of *transforming nature*, illustrating Buell's observation that "eloquent nature writing is not all there is to ecodiscourse."[42] Describing, for instance, the great port of Buenos Aires on the banks of the Riachuelo, Zeballos denounces its uselessness as a murky, shallow stream that, upon being deepened by the engineer Luis A. Huergo, promises to be transformed into "el Clyde argentino" (the Argentine Clyde)—that is, akin to Scotland's third longest river, which imperial Britain artificially straightened, widened, and deepened to increase commerce.[43] Zeballos thus engineers a sublime that is literary in its Humboldtian project but that is literal in its willingness to carve into the space of nature to instrumentalize it.

Proto-ecological Zeballos: Anticolonial Aesthetics

The term "environment," Buell tells us, did not enter English usage until the 1830s.[44] And the word "ecology" appeared some decades later in 1872.[45] Yet, as Ricardo A. Gutiérrez and Fernando J. Isuani trace in their reconstruction of Argentine environmentalism, more than a century passed before Great Britain created the position of Environmental Secretary in 1970. The United States, likewise, launched the Environmental Protection Agency in 1971. The next year Argentina followed in suit with the Asociación Argentina de Ecología, and on November 8, 1972, the engineer Eduardo Pous Peña composed a communiqué titled "El Dr. Estanislao S. Zeballos, Guardián argentino de los recursos naturales" (Argentine guardian of natural resources).

Peña's mid-twentieth-century tribute to nineteenth-century environmentalism *avante la lettre* historicizes the movement while showcasing

Zeballos's prophetic observations on deforestation and his aspirations to protect endangered species; to be sure, he avoids any mention of Zeballos's role in facilitating the extinction of an entire human ethnic group. By including the Argentine statesman among the nation's first conservationists, Peña situates Zeballos as a precursor to the international ecological movement taking place at the moment of his writing in the 1970s. He paints the portrait of a *prócer* who locates national progress at the intersection "la sensación estética," "un espíritu conservacionista," and a "sentido de economía" (aesthetic feeling, a conservationist spirit, and a sense of economy) and, without stating it explicitly, alludes to the didactic impulse undergirding Zeballos's aesthetic project.[46] For Peña, Zeballos's chief contributions as a pioneering and presaging conservationist were economic, while the Argentine nation's most regrettable error was to ignore his warnings in both realms.

In the remainder of this chapter, I will build on Peña's adroit but preliminary reflections to illuminate the ways in which Zeballos's literary register suggests a valorization of the natural world unseen in his earlier works. In addition to reframing nature so it is more *natural* than *natural resources* and, in turn, portraying Amerindians as more human than labor, Zeballos's narrative strategies point to ecodegradation's disproportionate consequences for already marginalized communities—indeed, a sort of early environmental justice that stands in stark opposition to his belief that progress would necessarily entail casualties along the way.

This dialectical tug-of-war lies at the heart of Zeballos's writings that most obviously present themselves as literature, namely his trilogy of Indianist novels *Callvucurá y la dinastía de los Piedra* (1884), *Painé y la dinastia de los zorros* (1886), *and Relmú, reina de los pinares* (1887), which are set between the 1830s–40s and are unequivocally bad: they drag with content that reproduces and even directly footnotes his earlier travelogues; their plots unfold as poorly conceived romances that exemplify Doris Sommer's famous formula for Latin America's foundational fictions, which posit national consolidation through the marriage of Eros and Polis; and their generic amalgamation satisfies neither those who seek history nor those who prefer literature.

For my purposes, the novels befuddle for a more important reason: their underlying contradiction. Zeballos immortalizes the very Indians whose death he ensured in what Balderston calls "nostalgic fictions which evoke a vanished world."[47] This evocation is particularly intriguing given the economic and social context informing coeval Argentine fiction, in which his

Indianist tales sat as outliers amidst the rise of copious gaucho and realist novels, works like *Arturo Sierra* by Julio Llanos or *La gran aldea* by Lucio Vincente López. Both bestsellers in their moment, these novels—published in 1884, the same year as *Callvucurá*—unfolded within the arc of extraordinary economic growth in 1880s Argentina and captured, respectively, both the promise of *ranchificación* of the rural Pampa and the social evolution of urban Buenos Aires.[48] I mention these works to highlight that not only did Zeballos receive little acclaim for his first novel but that he then astonishingly defied the literary market's desire to consume narratives depicting the very territorial changes he had underwritten in *Conquista*. Instead, he wrote the second installment of his trilogy, *Painé*, which appeared serially in *La Prensa* from June 15 to July 10, 1886. Against the backdrop of Argentina's dizzyingly fast steps forward with increased European immigration, commerce, and agriculture, Zeballos took two steps back to literarily preserve the very Amerindians whose annihilation he had coordinated.

While in the 1884 *Callvucurá* he continues to conjugate past and present to demonstrate progress—"*Los Rancúlches* ocupaban las privilegiadas tierras que son ahora rico teatro de especulaciones, de estancias y de pueblos" (The Rancúlches occupied privileged lands that are now the rich theatre of speculations, ranches and towns)—he simultaneously aligns natives with nature to valorize both.[49] Early in the novel, for instance, Zeballos portrays watersheds not in terms of utility but rather as the site of Indigenous revel and inebriation where the earth offers pine-nuts galore. The native place-name corresponds with nature's offering and enters his literary cartography, one now drafted from a horizontal plane rather than from above, one that immortalizes spaces as they were rather than what they have become, a place "en las márgenes del arroyo Cahuiñqué, donde se reunían periódicamente los indios para celebrar las grandes bacanales, cuyo arroyo fecunda la vega de Carahué y es hoy conocido por Pihuen ó 'de los pinos'" (on the margins of the Cahuiñqué stream, where the Indians gathered periodically to celebrate their grand bacchanals, whose stream fertilizes the meadow of the Carahué and is today known as Pihuen or "of the pinenuts").[50] Neither waterway nor agriculture nor plural subjectivities are in service of economic progress in this bucolic description, which alludes to long-gone notions of the inhabitant and of self-sufficiency. Zeballos's narration again betrays the ways in which imperialist nostalgia "conceals guilt," his florid prose glorifying an earlier epoch as if to obscure his complicity in vanquishing it.[51]

Similarly, in *Painé* (1886), Zeballos devotes Chapter 37 to "árboles frondosos" (lush trees) that are not felled to clear land for agriculture but that naturally offer protection from the elements, trees whose "copa redonda" (round dome) metaphorically opens "como paraguas colosales" (like colossal umbrellas).[52] These soldiers of the Pampas protect and nourish smaller flora and are effusively praised for their striking beauty, a treasure to all the senses as they "alimentaban una curiosa variedad de plantas inferiores, parásitos de flores hermosas y perfumadas . . . con bellísima variedad de formas y matices" (would feed a curious selection of smaller plants, beautiful, perfumed and parasitic flowers . . . with a gorgeous range of forms and shades).[53] Passages like these abound in the trilogy, none worth exploring in depth save to remark that Zeballos reveals himself over some 1500 pages to be increasingly attuned to nature's worth beyond its use-value, to be awakened and in fuller apprehension of species interdependence, and to be dialectically torn between celebrating his role in Argentina's spatial reconfiguration-cum-genocide and lamenting those consequential changes to the landscape.

I historicize Zeballos's Indianist trilogy to suggest that his novelistic idealization of frontier life and his romanticization of Amerindian populations may be explained by his increasing ecological consciousness, a conscious ness that does not only self-indict but also seems intermittently vexed with nineteenth-century capitalist projects informed by the trifecta of positivism, racism, and extractivism. In *Viaje*, for instance, he nimbly anatomizes the ways in which human activity causes measurable ecoharm by narrating a presumably veridical conversation between himself and Mr. John Brigest, an English *hacendado* from the southern plains. As they chat in a train car and take in the pastoral landscape—newly scarred with the very tracks on which they ride—Zeballos presents readers with a highly stylized dialogue in which the Englishman asks him questions about weather patterns and climate change. Mr. Brigest's tone is skeptical or even accusatory, recalling other Independence-era texts that include "scheming Englishmen."[54] Zeballos answers him with patience and erudition. This mode of didactic delivery informs both the Englishman and Zeballos's readers about Argentina's physical environment while bestowing Zeballos with authority.

The first question relates to floods: "¿Qué opina Vd. Doctor, dijo el ingles, sobre la causa de las inundaciones?" (What do you think, Doctor, said the Englishman, about the cause of the floods?) to which Zeballos responds, "Las inundaciones son producidas, señor, por órdenes de causas: 1° Las pendientes continental y rejionales del terreno. 2° La transformacion de la vegetacion" (The floods are produced, sir, for a few reasons: 1. The regional and continental shelves. 2. The transformation of the vegetation).[55] With this pointed enumeration he attributes climatic events to human activity, acknowledging and honoring the vast damage done to the terrain in the name of progress. When the Englishman pushes against his answer incredulously, Zeballos elects a different mode of delivery by shifting from words to maps: "Mi interlocutor parecia vivamente interesado en oirme y deseoso de corresponder á tal deferencia, abrí una maleta y desarrollé un mapa de Buenos Aires" (My interlocutor seemed intently interested in hearing me, and hoping to acquiesce to such deference, I opened a briefcase and sketched a map of Buenos Aires).[56] Some years later, Zeballos would employ the same strategy with Theodore Roosevelt to visually affirm his expertise by whipping out a map of the Argentine territory.[57] In both instances, he stakes claim to local knowledge and relocates imperial states to a position of deference, reorienting the axes of knowledge production southward.

Over the course of the seven-page dialogue, Zeballos signals an inchoate awareness that human hands have intensified floods and droughts, winds, and hurricanes, and that he should do something about it. According to Zeballos (who ventriloquizes the words of the *hacendado*), even the Englishman believes that the Argentine—a journalist, a scientist, and a statesman—has the responsibility to disseminate his findings and agitate opinion.[58] Yet paradoxically, just lines later, Zeballos insists that the very human activity he condemns is essential: "Hay materia inerte: oro en las arenas, diamantes en las montañas; pero materia siempre inválida, mientras la mano del hombre no le imprima movimiento, transformacion, valores" (There is inert matter: gold in the sand, diamond in the mountains; but matter is worthless unless man's hand brings about its movement, transformation, values).[59]

How may we explain Zeballos's about-face?

Zeballos's brand of social ecology reveals the difficulty of environmentalism in formerly colonized territories that seek the same extractive profits acquired by their colonizers. In this case, he appears to point a subtle finger at British neocolonial pillaging—metonymically represented by Mr. Brigest—by maligning Spain. As Jennifer French has shown, this tripartite structure recurs in the discourse of informal imperialism, which departs from the Manichean dyads of colonizer and colonized in favor of a triad: "The land is controlled by Latin American elites; labor is provided by the exploited populations of their countries; and the capital is predominantly British."[60] With capital on the line, any critique of the British neocolonial apparatus had to be divvied out indirectly and with great care.[61] In keeping with the figure of three and with an ambivalently anticolonial sensibility, then, Zeballos first incriminates Spain for failing to sufficiently extract from the New World's fertile lands, searching instead for mythical sites of riches like El Dorado and Jauja. By failing to cultivate nature's riches, Zeballos complains, Spain delivered nature back to nature, a tribute of sorts that returns to the bowels of the earth in the flows of lava.[62]

On the other end of the spectrum—a difference of degree and not of kind—England's informal empire sustains an equally detrimental hyperextractive economy. Thus on the heels of his condemnation, Zeballos's dialogue with Mr. Brigest turns to ravaging effects of overexploitation and the need to recalibrate Argentina's rural production, which he describes as "broken," be it because of flooding or drought that wreak havoc upon the land's surface, waterways, and animal life.[63] The lack of human attention to ecological catastrophe results in suffering at the lowest echelons, Zeballos presciently notes. Writing as a functionary of the state, he blames the State itself for perpetuating the devastation of lands and thereby the destitution of non-state actors who disproportionately suffer the consequences of untenable land use, describing it as a matter of poverty:

> No se siente con acaso toda la intensidad en las populosas capitales, porque el hacendado que pierde la mitad de sus vacas y tiene vastos terrenos en que salvar el resto, queda generalmente en pié; pero la mayoria es constituida en pequeños propietarios, dueños de reducido terreno y de corto número de animales, y son estos los que pierden el todo en aquellos desequilibrios pavorosos, que parecen escapar á la acción de los hombres, que los devotos atribuirán á azotes de los cielos, y que yo imputo á la falta de seriedad de nuestro carácter y á la falta no menos deporable y casi completa de preparacion administrativa, en todos ó en la mayor parte de los hombres públicos del pais.[64]

> It is hardly felt with the same intensity in the populous capitals, because the *hacendado* that loses half his cows and has vast terrains on which to save the rest, generally lands on his feet; but the majority consists of small proprietors, owners of less terrain and a small number of animals, and they are the ones who lose everything in those horrific imbalances, which seem to escape the action of men, which the devout attribute to a heavenly scourge, and which I impute to the lack of seriousness of our character and the

almost complete and no less deplorable lack of administrative preparation, among all or the majority of the country's public men.

The State is complicit and in cahoots with the English. Zeballos cannot make this accusation outright, but the subtext of his dialogue with the Englishman—an *hacendado* like the one Zeballos describes, who can lose half his livestock and still land on his feet—implicates elites of all stripes for their collusion, Argentine and British, be they landowners or businessmen or statesmen. Humanizing the science of climate change, Zeballos in all his enlightened rationality signals to the devout that their God is not meting out natural disasters; rather, he blames public officials who lack the character and preparation to take such threats seriously since they, and their most privileged constituents, do not experience the devastation as gravely.[65]

If we connect Zeballos's earlier accusations of the colonial apparatus with this defense of the destitute, then we might conclude that despite—and because of—his very participation, he understands that colonial *and* neocolonial projects contribute to social and environmental evils that predominate in formerly occupied territories. At the peak of his positive environmental unconscious, Zeballos's words reveal early awareness that the repercussions of "natural events are often distributed according to the tragedies of human oppression and poverty, and for precisely that reason, ecology cannot be ignored in the Global South."[66] He thus does not ignore ecology but rather faces it head-on, posing to his English interlocutor a clarion call that links capitalist production with climate-related devastation: "Necesitamos regularizar nuestra producción . . . : porque esta es para mi, Mr. Brigest, cuestion de clima. ¿Cómo se modifican, en efecto, las condiciones de fecundidad de un territorio y de que manera se regularizan?" (We must regulate our production . . . : because this is for me, Mr. Brigest, a question of climate. How do the conditions of fecundity of a territory change, in effect, and how do they become regulated?)[67] Caught somewhere between apostrophe and aversion, Zeballos's insertion of Mr. Brigest's name addresses a silent listener, a reading public to whom he poses the million-dollar question: how do we sustainably cultivate land *and* prevent cataclysmic disasters?

Zeballos's foresight prefigures contemporary efforts to theorize sustainable land use through a variety of disciplines, be they humanistic or scientific. He recognizes and critiques what today might be called, in Jyotirmaya Tripathy's terms, "Third World environmentalism," in which the "postcolonial state legitimizes its resolve to over-exploit the environment through the narrative of development," which immunizes it from both local and global criticism.[68] Tripathy notes that the coupling of land and livelihood banishes the Global South to a "pre-political condition" while confining any "activism to the pre-history of environmental consciousness."[69] Years before such debates were in vogue, Zeballos inserts Argentina into both the politics and the history of environmental consciousness. To close the dialogue with Mr. Brigest, he describes massive territorial transformation, signaling that the nation's terrain has experienced radical changes over the course of a half-century largely because the "viejos *pajonales*" (old scrublands) have been replaced by agricultural fields that invade the territory in every direction.[70] This invasion leads to the doomed marriage of increased evaporation and decreased absorption, such that in the rainy season "se pierde en vapores una gran parte de agua, que debiera profundizar el humus y el resto produce la inundacion por falta de declives y desagues" (a great deal of water evaporates rather than going deep into the soil and the rest produces floods due to a dearth of incline and drainage).[71]

The problem succinctly articulated, Zeballos offers a solution that is economic at its heart: "volver á la tierra su abrigo protector y su absorvente esponjosidad" (return to the earth its protective cover and absorbent sponginess) by ensuring that "los gobiernos tomaran á pecho la plantacion de arboledas (governments seriously consider planting trees). Vacas y cueros, ovejas y lanas, campos y pastos, valdrían cinco veces mas para posperidad del País" (Cows and leather, sheep and wool, fields and pastures, would be worth five times more for the prosperity of the country).[72] Likely indebted to his reading of Humboldt—who as early as 1801 blamed deforestation for the desiccated basin around Venezuela's slowly sinking Lago de Valencia—Zeballos's remedy to cure the earth's ills was no ersatz science.[73] Rather, replanting ravaged lands has long since been proven to restore balance to ecosystems and prevent natural calamities. His environmental unconscious awakened to human interdependence with nature, Zeballos the proto-ecologist and genocidal extractivist put forth an idea in 1881 that he would manifest two decades later with the inaugural Día del Árbol in 1901. We thus come full circle, the answer affirmative—and ambiguous—to Buell's closing question in *Writing for an Endangered World*: "Will any act of environmental imagination achieve the kind of hearing an environmentalist would wish?"[74]

NOTES

1. Cernadas et al., *Escenarios de la sociabilidad*, 31.
2. Viñas, *Indios, ejército y frontera*; Andermann, *The Optic of the State*; William Henry Hudson, *Idle Days in Patagonia*.

3. Buell, *Writing for an Endangered World*, 9.
4. Buell, *Writing for an Endangered World*, 4.
5. "Siluéticas políticas," *La Nota* 26, 501; Balderston, "The Indianist Novels," 323.
6. Rodríguez, *Relics & Selves*.
7. Sarmiento, "Quince mil leguas," 82.
8. Scott, *Seeing Like a State*, 13.
9. Navarro, *Scribere est agere*, 27.
10. Peña, "El Dr. Estanislao S. Zeballos," 43–54.
11. Suárez, "La región del Trigo (¡y el agua!)."
12. Zeballos, *La conquista de quince mil leguas*, 49.
13. Marzec, "Speaking Before the Environment," 424.
14. Buell, *Writing for an Endangered World*, 21–22. Drawing on Leslie Silko's observations on the differences between Indigenous and non-Indigenous landscape representations, Buell notes that "particularity of environmental detail may actually betoken lack of connectedness" (21).
15. Buell, *Writing for an Endangered World*, 21.
16. Zeballos, *La conquista de quince mil leguas*, 240.
17. Zeballos, *La conquista de quince mil leguas*, 91.
18. Zeballos, *La conquista de quince mil leguas*, 90.
19. Scott, *Seeing Like a State*, 2; Pratt, *Imperial Eyes*, 201; Spurr, *The Rhetoric of Empire*, 17; Taylor, "Remapping Genre," 1,420.
20. Scott, *Seeing Like a State*, 2.
21. Uriarte, *The Desertmakers*, 145; Marzec, "Speaking Before the Environment," 423.
22. Marzec, "Speaking Before the Environment," 423.
23. Thornber, *Ecoambiguity*, 5. This section's title borrows from Karen Thornber's notion of *ecoambiguity*, which aims to make meaning of contradictory encounters between humans and ecosystems, particularly as they manifest in creative writing.
24. Súarez, "La región del Trigo (¡y el agua!)," 1.
25. Rosaldo, "Imperialist Nostalgia," 108.
26. Zeballos, *La rejion del trigo*, 26.
27. Zeballos, *La rejion del trigo*, 27.
28. Zeballos, *La rejion del trigo*, 27.
29. Zeballos, *La rejion del trigo*, 27.
30. Buell, *Writing for an Endangered Word*, 18.
31. Buell, *Writing for an Endangered Word*, 19.
32. Buell, *Writing for an Endangered Word*, 16.
33. Zeballos, *La rejion del trigo*, 27.
34. Zeballos, *La rejion del trigo*, 27.
35. Zeballos, *La rejion del trigo*, 27.
36. See Chapter 2 of Madan, *Lines of Geography in Latin American Narrative*.
37. Zeballos, *Viaje al país*, 9.
38. Zeballos, *Viaje al país*, v.
39. Zeballos, *Viaje al país*, vi.
40. Zeballos, *Viaje al país*, vi.
41. Zeballos, *Viaje al país*, vi.
42. Buell, *Writing for an Endangered World*, 195.

43. Zeballos, *Viaje al país*, 12.
44. Buell, *Writing for an Endangered World*, 2.
45. White Jr., "The Historical Roots of Our Ecological Crisis," 5.
46. Peña, "El Dr. Estanislao S. Zeballos," 50–54.
47. Balderston, "The Indianist Novels," 324.
48. Lichtblau, *The Argentine Novel*, 134–44.
49. Zeballos, *Callvucurá*, 17.
50. Zeballos, *Callvucurá*, 8.
51. Rosaldo, "Imperialist Nostalgia," 109.
52. Zeballos, *Painé y la dinastía*, 98.
53. Zeballos, *Painé y la dinastía*, 100.
54. French, *Nature, Neo-Colonialism*, 6.
55. Zeballos, *Viaje al país*, 17–18.
56. Zeballos, *Viaje al país*, 18.
57. Zeballos, "Discurso del Doctor Estanislao Zeballos," 499.
58. Zeballos, *Viaje al país*, 20.
59. Zeballos, *Viaje al país*, 20.
60. French, *Nature, Neo-Colonialism*, 155–56.
61. French, *Nature, Neo-Colonialism*, 5–7.
62. Zeballos, *Viaje al país*, 21.
63. Zeballos, *Viaje al país*, 21.
64. Zeballos, *Viaje al país*, 21.
65. Zeballos, *Viaje al país*, 21.
66. Handley, "Down Under," 95.
67. Zeballos, *Viaje al país*, 22.
68. Tripathy, "Indian Environmentalism," 72.
69. Tripathy, "Indian Environmentalism," 73.
70. Zeballos, *Viaje al país*, 22.
71. Zeballos, *Viaje al país*, 22.
72. Zeballos, *Viaje al país*, 22.
73. Humboldt, *Personal Narrative*, 42.
74. Buell, *Writing for an Endangered World*, 265.

BIBLIOGRAPHY

Andermann, Jens. *The Optic of the State: Visuality and Power in Argentina and Brazil*. University of Pittsburgh Press, 2007.

Balderston, Daniel. "The Indianist Novels of Estanislao S. Zeballos." *Revista Canadiense de Estudios Hispánicos* 15, no. 2 (Winter 1991): 323–27.

Buell, Lawrence. *Writing for an Endangered World*. Harvard University Press, 2001.

Cernadas, Mabel N., Lucía Bracamonte, María de las Nieves Agesta, and Yolanda de Paz Trueba. *Escenarios de la sociabilidad en el sudoeste bonerense durante la primera mitad del siglo XX*. Editorial de la Universidad Nacional del Sur, 2016.

French, Jennifer. *Nature, Neo-Colonialism, and the Spanish American Regional Writers*. Dartmouth College Press, 2005.

Gutiérrez, Ricardo A., and Fernando J. Isuani. "La emergencia de ambientalismo estatal y social en Argentina." *Revista de Administração Pública* 54, no. 2 (April 2014): 295–322.

Handley, George B. "Down Under: New World Literatures and Ecocriticism." *Global South* 1, no. 1 (Winter 2007): 91–97.

Humboldt, Alexander von. *Personal Narrative of Travels to the Equinoctial Regions of the New Continent During the Years 1799–1804*, vol. 4. Translated by Thomasina Ross. Bell & Daldy, 1871.

Hudson, William Henry. *Idle Days in Patagonia*. E. P. Dutton, 1917. Originally published in 1893 by Chapman and Hall.

Lichtblau, Myron. *The Argentine Novel in the Nineteenth Century*. Hispanic Institute in the United States, 1959.

Madan, Aarti S. *Lines of Geography in Latin American Narrative: National Territory, National Literature*. Palgrave, 2017.

Marzec, Robert P. "Speaking Before the Environment: Modern Fiction and the Environment." *Modern Fiction Studies* 55, no. 3 (Fall 2009): 419–42.

Navarro, Fernando "En busca de Zeballos," In *Scribere est agere: Estanislao Zeballos en la vorágine de la modernidad argentina*, 11–36. Edited by Sandra Fernández and Fernando Navarro. La Quinta Pata & Camino Ediciones, 2011.

Peña, Eduardo Pous. "El Dr. Estanislao S. Zeballos: Guardián Argentino de los recursos naturales." *Academia Nacional de Agronomía y Veterinaria* (November 1972): 43–54.

Pratt, Mary Louise. *Imperial Eyes: Travel Writing and Transculturation*. Routledge, 1992.

Rodríguez, Fermín. "Estanislao S. Zeballos: Un desierto para la nación." In *Relics & Selves: Iconographies of the National in Argentina, Brazil and Chile, 1880–1990*, edited by Jens Andermann and Patience A. Schell. Iberoamerican Museum of Visual Culture on the Web, 2005. http://www7.bbk.ac.uk/ibamuseum/texts/Rodriguez01.htm.

Rosaldo, Renato. "Imperialist Nostalgia." *Representations*, no. 26 (Spring 1989): 107–22.

Sarmiento, Domingo Faustino. "Quince mil leguas." In *Obras de D. F. Sarmiento*, edited by A. Belin Sarmiento. Mariano Moreno, 1900.

Scott, James C. *Seeing Like a State: How Certain Schemes to Improve the Human Condition Have Failed*. Yale University Press, 1998.

"Siluéticas políticas." *La Nota* 26 (1916): 501–3.

Spurr, David. *The Rhetoric of Empire: Colonial Discourse in Journalism, Travel Writing, and Imperial Administration*. Duke University Press, 1993.

Suárez, Pablo Ernesto, "La región del Trigo (¡y el agua!)." Instituto en Formación de Recursos Hídricos, October 2016. https://www.ina.gob.ar/ifrh-2016/trabajos/IFRH_2016_paper_49.pdf.

Taylor, Diana. "Remapping Genre Through Performance: From 'American' to 'Hemispheric' Studies." *PMLA* 122, no. 5 (October 2007): 1,416–30.

Thornber, Karen Laura. *Ecoambiguity: Environmental Crises and East Asian Literatures*. University of Michigan Press, 2012.

Tripathy, Jyotirmaya. "Indian Environmentalism and its Fragments." In *Ecoambiguity, Community, and Development: Toward a Politicized Ecocriticism*, edited by Scott Slovic, Swarnalata Rangarajan, and Vidya Sarveswaran, 71–84. Lexington Books, 2014.

Viñas, David. *Indios, ejército y frontera*. Siglo Veintiuno, 1982.

White Jr, Lynn. "The Historical Roots of Our Ecological Crisis." In *The Eco-Criticism Reader*, edited by Cheryll Glotfelty and Harold Fromm, 3–14. University of Georgia Press, 1996.

Uriarte, Javier. *The Desertmakers: Travel, War, and the State in Latin America*. Routledge, 2019.

Zeballos, Estanislao Severos. *La conquista de quince mil leguas*. La Prensa, 1878.

Zeballos, Estanislao Severos. *Viaje al país de los araucanos*. Jacobo Peuser, 1881.

Zeballos, Estanislao Severos. *La rejion del trigo*. Jacobo Peuser, 1883.

Zeballos, Estanislao Severos. *Callvucurá y la dintastía de los Piedra*. 1884. Reprint, Recuerdos Argentinos, 1890.

Zeballos, Estanislao Severos. *Painé y la dinastía de los zorros*. 1886. Reprint, J. Peuser, 1889.

Zeballos, Estanislao Severos. "Discurso del Doctor Estanislao Zeballos en la Universidad de Buenos Aires. El 10 de noviembre." *Boletín Mensual del Museo Social Argentino* 2 (1913): 483–500.

CHAPTER 9

La Revista Hispano-Americana (1895–1896)

Laura Méndez's Extractivist Pedagogy

CATALINA RODRÍGUEZ

In the short story "La curva" (1908), the Mexican writer Laura Méndez (1853–1928) recounts a ruse. Silverio Madariaga, the main character of the story, sells his profitable land in California below market value because he is unable to read either English or Spanish and is openly unwilling to entertain the contents of the newspaper. His tenant Mr. Wilson, after offering to teach Silverio how to read and accepting his refusal, buys the land just in time to profit from the new railroad. The swindle is thus mobilized by a combination of Mr. Wilson's astuteness and Silverio's lack of interest in education. Aside from recounting the ruse, "La curva" also uses its main characters to present contrasting relationships between humans and nature. On the one hand, the Mexican American Silverio experiences the land and the house as an extension of his own body. When Wilson demolishes the house Silverio feels "los golpes de barreta en todo su ser . . . [y] le pareció que los componentes de todo su ser se disgregaban y desaparecían dispersos por el espacio" (the pry bar blows in his whole being . . . [and] he believed that the components of his being came apart and disappeared, scattered in space).[1] On the other hand, the story uses the Anglo-American Wilson to represent an instrumental relation between nature and humans. For Wilson the land is only a potential source of wealth and progress: "lo que los Americanos debían hacer era comprarlo todo; el río, el valle, el monte, seguros de que, a vuelta de cinco años, cada palmo de tierra valdría diez veces su precio actual" (what the Americans had to do was buy everything; the river, the valley, the mountain, confident that in five years every inch of land will be worth ten times its current price).[2] The two approaches to the land work as signifiers of irreconcilable differences between Mexicans and Anglo-Americans, as well as the tension between affective and instrumental relations to the land. At the end of the story Silverio regrets his lack of interest in education and begins to view Wilson's relationship with the land as superior. In this regard, the swindle in "La curva" mobilizes a cautionary tale that aims to warn Mexicans and Latin Americans of the disadvantages they face when they privilege ignorance over education and sentimental attachment over financial gain. This becomes even clearer by the end of the story, when Silverio promises a doll to his daughter as a reward if she learns how to read and write fluently.

The narrative arc of "La curva" is one example of how Laura Méndez endorsed an instrumental relationship between humans and nonhuman nature in an attempt to address the disparities between Latin- and Anglo-Americans in the aftermath of the US Intervention in Mexico (1846–1848). The story is also representative of Méndez's broader concern for strengthening education and fostering economic development. This comes as little surprise given that Méndez devoted her life to the Mexican education system, where she held positions as teacher, director, and school inspector.[3] As Mílada Bazant points out, Méndez was one of the most prominent teachers of turn-of-the-century Mexico, and her opinion was highly regarded within the political spheres of the Porfirio Díaz regime. Between 1903 and 1908 Diaz himself commissioned Méndez to observe the education systems of the United States and Germany.[4] Her role as an educator not only determined her reception in Mexican historiography but it also framed her fictional and journalistic voice. When she published "La curva," in 1908, Méndez turned to the fictional realm to highlight the scarcity of education not only within Mexico but also among Mexican Americans living in California.[5] Education and a new understanding of the potential wealth associated with fertile lands appear as the only possible solution to Silverio's story, and as a solution that, as we see by the end of the story, can only be pursued by Silverio's daughter.

An investment in educating readers about the potential wealth of the land is present throughout Méndez's oeuvre, especially the bilingual economic magazine *La Revista Hispano-Americana* (1895–1896), which Méndez founded and directed from San Francisco. Like "La curva," the magazine demonstrates Méndez's desire to address economic and cultural disparities between the United States and Latin America by encouraging a new kind of relationship between Latin Americans and their land. Despite its evident success—between advertising and subscriptions, *La Revista* recorded monthly profits of $1,000 dollars—and its position as one of the first bilingual economic magazines directed by a woman, *La Revista Hispano-Americana* remains understudied in the fields of Latin American and Latinx scholarship.[6] Recent work on Méndez has focused on her role within Mexico's education system as well as her contributions to feminist thought in the early twentieth century, while critical studies on Méndez's literary writing have centered on her poetry, her novel *El espejo de Amarilis* (1902), and the short story collection *Simplezas* (1910), in which "La curva" was included.[7] *La Revista* was concerned mainly with strengthening market trade and providing economic advice, and as such appears to be very different from Méndez's work as fiction-writer and educator. If this difference in genre and medium helps to explain *La Revista*'s absence from scholarly examinations of Méndez's work, these apparent differences will in fact enable us to better understand just how fundamental the goal of fomenting an instrumental relation between humans and nature was to Mendez as an early feminist and an elite Latin American intellectual.

In this chapter, I turn to *La Revista* to uncover how Méndez's pedagogical commitments coincide with the rhetorical strategies used by the magazine to call for extraction and industrial development in Latin America. The magazine advanced a pedagogical initiative that relied on Western notions of civilization, progress, and education to justify the extraction and sale of natural resources; what is more, *La Revista* takes advantage of its bilingual approach to inscribe two different goals. For its English readers the magazine was a catalogue of investment possibilities in Latin America, while for its Spanish audience, *La Revista* attempted to equalize dynamics of extraction and development through pedagogy. In some ways, *La Revista*'s double discourse was promoting, or at least risking, hierarchical relations between north and south because it encouraged Anglo-American capitalists to understand Latin American resources as "ripe" for development.[8] I take this rhetorical strategy as evidence of what I call Méndez's "extractivist pedagogy," an initiative that aimed to teach elite Latin American readers that exploiting the land was the first and most important step toward progress and civilization. Building on the work of Eduardo Gudynas I understand extractivism as the appropriation of natural resources in mass, at exponential rates and through processes that alter environmental conditions.[9] Despite the fact that the concept was coined several years after Méndez's *Revista* appeared, it can help us understand how the dynamics of what would be later called extractivism were being promoted through discourses of modernization and progress since the late nineteenth century. The magazine allowed Méndez to combine her pedagogical interests with a promotion of a new relation between humans and the not human nature.

To promote an "extractivist pedagogy" *La Revista* adopted the same discourse that Europeans had created to describe America since the sixteenth century, in Mary Louise Pratt's words: "What held for Columbus held again for Humboldt: the state of primal nature is brought into being as a state in relation to the prospect of transformative intervention from Europe."[10] Latin America appears in *La Revista* mainly as a recognizable and categorizable nature whose potential wealth awaits human intervention. However, the magazine was not only interested in replicating the canonical discourse around Latin America as a region too "primitive" to take advantage of its own resources. Instead, and in part because it was produced in what Pratt calls a "contact zone," the magazine intended to alter the actors and hierarchies of exchange.[11] Thus Méndez's "extractivist pedagogy" encouraged elite Latin Americans to appropriate the dynamics of extraction that had been executed to their disadvantage and subjection. Therefore, the overarching message of *La Revista* was very similar to what Silverio understood at the end of "La curva," to wit: that knowledge and education could shed light on the instrumentality of nature and consequently help equalize international economic relations.

Learning from the Neighbor

Méndez established *La Revista Hispano-Americana* in February 1895, just six years after arriving in San Francisco. The writer had decided to start a new life on the other side of the US-Mexico border following the death of her husband, the poet Agustín F. Cuenca (1850–1884). In San Francisco, Méndez learned English, taught Spanish to wealthy women, and eventually made a place for herself in the world of journalism by founding *La Revista* with Argentine consul José Schneiden. The periodical's offices were headquartered in the prestigious Mills Building, which allowed Mendez to build connections with owners of newspaper emporiums and preeminent journalists.[12] *La Revista* was a very successful enterprise that quickly gained a large number of subscriptions in the Americas, especially among elite Latin Americans. In Mendez's own words: "De cuatro mil ejemplares que tiramos vendemos unos 30 en San Francisco, cosa de 120 en el resto del país y lo demás en Centro y Sud América" (From four thousand issues published, we sell around 30 in San Francisco, another 120 within the country, and the rest in Central and South America).[13] The magazine had a transnational audience that reached from the United States to the Pacific coast of Chile. Despite being advertised as a bilingual enterprise, the majority of the articles were published in Spanish. This suggests that *La Revista*'s main intended audience was a transnational Latin American community, one that could presumably understand both languages, and that the articles catering to English-speaking readers were secondary. The magazine's transnational audience was precisely what allowed for the endorsement of a pedagogical discourse aimed at Latin America as a whole.

However, the bilingual approach used by *La Revista* was pedagogically important because it created a dual rhetorical strategy. The English sections promoted new market trade and investment opportunities between the United States and Latin America while at the same time showing Latin Americans how Anglo-Americans understood land and nature.[14] The Spanish sections combined the promotion of new investment possibilities in Latin America with articles that simultaneously praised progress and economic development in the United States and Europe. The English sections were not solely intended for the Anglo-American readers because, as we've seen, they comprised a very small portion of *La Revista*'s audience. Instead, I propose, both English and Spanish discourses were crucial for the magazine's extractivist pedagogy insofar as they contained different rhetorical strategies that aimed to educate elite Latin Americans in how to take advantage of the potential wealth of their lands. The Spanish sections further exemplified the consequences of the extractivist model by including articles that praised California's success in economic, cultural, and technological terms.[15] The articles talked about industrial developments, agricultural successes, urban planning, new buildings, railroads and libraries, among other things. The editors described California as a paragon of civilization and modernity, a characterization apparent even in their selected titles (e.g.,

"California Awakes," "A valuable library," and "To be applauded").[16] Each issue combined those articles on California with chronicles about different Latin American regions that repeatedly emphasized the lack of human intervention. Through the combination of both discourses, a constant praise of California and a commentary on Latin America's potential development, *La Revista* constructed an example of what Latin America could be if it were to follow in the steps of California. Despite the fact that California had been part of Mexico until 1848, just forty-seven years earlier, the magazine never mentions its connection to the country or to the Latin American region. This was a concern that Méndez elaborated in other articles and short stories and a very important aspect of "La curva."[17] The fact that *La Revista* does not mention California's past as part of Mexico, however, does not mean that its readers were not aware of it. On the contrary, the audience was likely aware of the recent historic change, and this was precisely what made California the perfect example. *La Revista* was ultimately praising the changes that the Anglo-American extractivist model achieved in a territory that, not long before, was at the same stage as northern Mexico.

An article from the first issue, "California y sus elementos de vida" (California and its life elements), is a good example of how the rhetoric operated. It starts by emphasizing the state's history of successful gold mining: "la industria minera . . . ha alcanzado la plenitud de su desarrollo merced a los notables inventos de maquinaria" (the mining industry has reached the peak of its development thanks to notable inventions in machinery).[18] The first two paragraphs portray California's mining industry as an example of technological development, one that had reached its peak. From mining, the article moves to present California's emergent agricultural industry and praise its development and modernization:

> el éxito no se hizo esperar y hoy por hoy, California, empezó a cobrar empuje llegando a un vigor gigantesco tanto en su producción agrícola como en su desarrollo comercial. . . . A este rápido desarrollo ha respondido necesariamente el crecimiento de las ciudades existentes, la creación de otras nuevas, el embellecimiento de todo género de edificios, los cuales de carpas o casuchas miserables de madera, que eran en 1849, han sido transformados en palacios de ladrillo, granito, mármol, hierro y madera.

> Success came swiftly and today, California has started to gain momentum and has achieved a gigantic force both in its agricultural and commercial development. . . . Which has resulted in the growth of existing cities and

the creation of new ones, the beautification of every kind of building. Buildings that in 1849 were tents and wood shacks have now been transformed into brick, granite, marble and iron palaces.[19]

Agriculture appears as a solution to gold exhaustion, one that is also a natural consequence of the fertility and abundance of the land. The agricultural industry is named only as a gateway to praise California's development and success. The article even emphasizes how the "humble shacks from 1849" have been transformed into "marble, brick and iron palaces."[20] The date is very important because it brings the readers back to California's first year as part of the United States and marks the beginning of the modernization. In other words, the article aims to show that civilization and development are a consequence of the strong mining and agricultural industries that the Anglo-Americans were able to create.

Interestingly enough, the descriptive rhetoric employed to represent California's land is replicated throughout the magazine when different Latin American countries are discussed. The only difference lies in how the description of Latin American countries is never followed by a commentary on the development of buildings and cities. In the second issue of March 1895, for instance, the editors published a short Spanish-language article that summarized the characteristics of the Cauca region in Colombia. The article described geographic and environmental conditions in great detail, including area measurements, population, and average temperatures. Unlike the article on California, however, this imagery is not followed by a commentary on the success of human labor. Instead the authors emphasize the lack of human intervention through strongly gendered language: "la Naturaleza se presenta allí espléndida y solo esperando que la mano del hombre vaya a recoger y a aprovechar los frutos con que ella le brinda" (Nature presents herself splendidly and only waits for the hand of man to collect and take advantage of the fruits she offers).[21] The comparison between the two articles make evident how California was envisioned as a model of development and modernization for Latin America. If the agricultural industry has transformed California's humble shacks into palaces, the fertile and beautiful region of Cauca is just waiting for its turn. The intention was precisely to make elite Latin Americans aware of the potential wealth and development that existed in their lands, and—for better or worse—to show them the advantages of the Anglo-American model of economic development.

Throughout the rest of its issues *La Revista* maintains a comparison between California's recent development and Latin American's primitive exploitation. Always in Spanish, and always emphasizing contrast, the discourse adopts a clear pedagogical intention to show what an instrumental relation between land and humans could achieve in terms of progress and civilization. California's development appears as the selected example because its recent history shows that understanding land and nature as sources of wealth can have a strong consequence in the progress of a region. Further, California's recent past as a state of northern Mexico makes it the perfect example of how fifty years under an extractivist model can have a radical impact in the development of urban areas.

A Latin American Alliance

In order to educate the Latin American region as a whole, *La Revista* needed to strengthen the idea of the region as a homogenous "imagined community" that could replicate California's example.[22] This intention was made apparent from the first issue of the magazine with an article titled "Nuestra posición en la prensa" (Our position in the press):

> La favorable acogida que el público en general ha dispensado a nuestra publicación, nos demuestra que no estábamos equivocados al suponer que la creación de un periódico de información, que estrechara las relaciones mercantiles de California con las repúblicas latino-americanas respondería a una necesidad urgente. ¿Y qué mejor manera de afianzar los vínculos entre los países, que comunicándolos entre sí por medio de referencias que les permiten conocerse recíprocamente? . . . Nuestro programa: cooperar al desarrollo de las repúblicas de México, América Central y del Sur, haciendo que ellas entre sí se comuniquen y se estrechen como procedentes de una misma raza, y que todas juntas entablen con California, el progresista *Estado de Oro*, una serie de no interrumpidas transacciones comerciales.

> The favorable reception of our publication shows that we were not mistaken to suppose that the creation of an informational periodical that could tighten mercantile relations between California and the Latin-American nations was urgent. What could better consolidate connections between countries than allowing them to know and understand each other? . . . Our program: to cooperate with the development of Mexico, Central

America and South America, allowing those nations to communicate and strengthen their relations as part of the same race, and, also to establish commercial transactions between those nations and California, the progressive Golden State.[23]

The article summarized *La Revista*'s main intentions and emphasized the need to foster identification among Latin American nations. The rhetorical question encapsulated the magazine's fundamental assumption: that Latin American countries would form a community once they became better informed about each other. From the first issue on, the editors follow the intention outlined and include articles that provide information about each Latin American country, from its natural resources and geographical particularities to legislators, famous writers, and development projects. The magazine employs similar language to represent Guatemala, Colombia, El Salvador, Perú, Costa Rica, Argentina, Brazil, and México, respectively. For example, when talking about the Cordoba region in Argentina, *La Revista* starts by saying, "Los valles Córdoba están muy bien regados por varios arroyuelos y producen toda clase de vegetación de tierras templadas" (The Cordoba valleys are very well watered by multiple streams and produce every single kind of vegetation from temperate lands), and finishes by assuring that the region has enough barren lands to accommodate "thousands of souls."[24] While describing the State of Mexico it asserts: "Toda clase de cereales se dan en las regiones. . . . Los bosques y las selvas son notables aún más por su riqueza que por su variedad" (All kinds of grains are present in the regions. . . . Both forests and jungles are more notable for their richness than their diversity), and then closes by emphasizing that what is missing are new industries.[25] The national differences of each country are dissipated by descriptions that focus on the potential wealth of the land while emphasizing two things: a notion of primitive exploitation and the need to establish new industries. Ultimately, the disparities between countries are only manifested through products that *La Revista* commodifies as examples of fertility and abundance. When talking about Argentina, the magazine emphasizes its vineyards; the articles about Costa Rica stress a large production of coffee; and the descriptions of Colombia and Brazil, highlight the production of gold.[26] The description of every Latin American nation creates a notion of shared experiences and needs by representing all national lands as fertile but still widely unexploited. In other words, the ties between nations and the potential to "strengthen relations as part of the same race" responded to the magazine's representation of Latin American nations under terms that had been common since the sixteenth century: an abundant wealth of natural resources lacking human intervention. Méndez employs the rhetoric as a strategy to reconcile each country's particularities and demonstrate the need to establish a regional community.

Further, the magazine emphasized a common origin, a universal Latin American race, that aimed to bundle heterogeneous national groups together. This effort fell into an act of erasure not only of national particularities but mostly of racial differences. The universalization of Latin Americans was common in the late nineteenth century and had important impacts on how the national identity of most countries will continue to obliterate racial identities and hierarchies.[27] The idea of a Latin American race responded to *La Revista*'s intention to promote a homogeneous community that could be educated and therefore inscribed into global economic trends. Therefore, *La Revista*'s Latin American community was dependent on the magazine's own representation of the region under similar terms.[28] *La Revista* was published just four years after the canonical "Nuestra América," in which José Martí, also writing from the United States, voiced similar intentions and concerns.[29] Therefore, Méndez's magazine could be understood as an example of late nineteenth century preoccupations with the formation of a Latin American community. *La Revista*'s discourse, crucially, leaves out the discussion of political institutions and forms of government and centers on how to fortify the new community economically. Méndez promotes an appropriation of the extractivist model precisely because her time in San Francisco makes her aware of the strategies and particularities behind the United States' rapid development. Instead of focusing on how to govern the Latin American nations, the magazine was concerned with how to make them economically successful while helping the elites profit from them. The magazine keeps the rhetoric of the abundant natural space but directs it at elite Latin Americans as a means to promote a change in their relation with the land and support the strengthening of a transnational community.

Feminism and Extractivist Pedagogy

Instructing Latin American readers in the advantages of the extractivist model relied on the reification of hierarchical dichotomies that confirmed the canonical opposition between mind/body, reason/feelings, human/non-human and male/female.[30] The magazine used gendered signifiers as a way to represent the land as abundant and everlasting. The feminization of the land was a key component of *La Revista*'s extractivist pedagogy. The article "California y sus elementos de vida" that we analyzed previously provides an example of how the gendered language was instrumentalized in the representation of the land: "una tierra acariciada pródigamente por el sol, [y] regada por importantes ríos . . . no podía menos que encerrarse un nuevo manantial de riqueza, si alguién había que quisiera depositar semillas en los surcos y aguardar pacientemente su germinación" (a land lavishly caressed by the sun, and irrigated by major rivers . . . could not enclose anything less than a new wealth spring, if there was someone to deposit seeds in furrows and patiently wait for the sprouting).[31] The land's description is gendered from the beginning, especially because it's described as passively waiting for someone to make it productive. Agriculture is represented through a set of images that denote reproduction and that center the land as a nubile woman waiting to be inseminated. The "alguien" that the article creates to assign the role of depositing the seeds is presumed to be male as a result of the land's feminization. This article is only one example of the rhetoric employed throughout the magazine to represent both land and nature.

The gendered language is also crucial for the description of Latin American lands. For instance, the chronicle titled "Colombia" describes the country as a "beautiful, fertile, progressive and rich" land that—like California prior to annexation by the US—is in need of exploitation.[32] The rhetoric of the article erases peoples and cultures, emphasizing instead natural resources as commodities. Colombia's "beauty" is directly linked to the size of its fertile lands and its potential wealth: "La soberbia cordillera andina que atraviesa el país en tres ramales, le da un aspecto montañoso, permitiendo en aquel privilegiado suelo todos los climas. . . . Como beneficiada por todos los climas, Colombia guarda en su seno el germen prepotente de millares de millones de plantas" (The superb Andean range traverses the country in three branches creating a mountainous aspect that, in turn, affords the privileged land every climate. . . . As it is blessed with every climate, Colombia holds in her breast the astounding seeds of a thousand millions of plants.)[33] In this description the notions of nurturing and motherhood are instrumentalized through the common Spanish phrasing "guarda en su seno." The commonality of the phrasing shows how the feminization of the land was a usual rhetoric strategy. The country's fertility is located in imagined breasts that are holding an outstanding number of seeds. The products that result from those seeds are largely described and the article closes with several paragraphs that list endemic trees, unique fruits and vegetables, medicinal plants, types of wood, pearls, tortoiseshells, minerals, especially ore, and gemstones. The notions of maternity and reproduction become key for the magazine's representation of the Colombian land, in this case, and of nature in general.

The prevalence of a gendered description of nature and the land in *La Revista* proves that the promotion of an instrumental relation between elite Latin Americans and their land was dependent on the ratification of gender difference. As Carolyn Merchant has shown, in Western cultures that distinctively separate the categories of mind and body, women are consistently associated with the supposedly "lower" order of nature and animality and men with the "higher" order of rationality.[34] Méndez's *Revista* is an example of how the use of female tropes to represent fertile and abundant lands is crucial in order to encourage progress, development, and modernization.

Beyond confirming the crucial role that the female/male dichotomy has had for relations between humans and nonhumans in the West, the reification of those gendered tropes in *La Revista* allows us to better understand Méndez's feminism. From a contemporary perspective, the reification of the gender binary that we see in the *Revista* seems to be in tension with Méndez's own role as an important early twentieth-century feminist. Méndez directed *La Mujer Mexicana* (1904–1906), was president of the Sociedad Protectora de la Mujer, and her work as editor of *La Revista* was even praised by the San Francisco women newspaper publishers for being "avant-garde because it departed from themes usually addressed in publications edited by women."[35] As Leticia Romero Chumacero highlights, Méndez's feminism followed in the footsteps of other Mexican writers such as Laurena Wright (1846–1896), who believed in the "civilizing power of education."[36] The writer understood feminism as a revolution of middle-class women who "quieren ser médicos, abogados, literatos, legisladores y cuanto hay, en vez de muñecas de tocador" (wanted to become doctors, lawyers, writers, legislators, and everything else instead of well-dressed dolls).[37] From her perspective, therefore, feminism was a fight to guarantee equal education for upper- and middle-class women. These ideas were shared by other contemporary Mexican feminists in *La Mujer Mexicana*, where writers made clear that they were not interested in advocating for woman suffrage but instead wanted to secure equal educational opportunities and paid work.[38] The movement adhered to hierarchical dichotomies and understood feminism as a tool to civilize the Mexican nation. Education was as important in Méndez's feminist endeavors as it had been in her efforts to establish an independent life and journalistic projects.

Méndez's feminism even justified an argument for women's education that still endorsed a traditional gendering of the private sphere: "Quién ha dicho que su verdadero puesto es el hogar, ha dicho muy bien; pero quien supone que para ocupar dignamente ese 'verdadero puesto' no ha menester sino tintura de los conocimientos humanos no tiene ni siquiera noción del significado moral de la familia" (Those who said that their true position was the home, said it rightly; but those who assume that, to respectably occupy this "true position," women don't need anything but a tinge of human knowledge, doesn't even understand the moral meaning of family).[39] Méndez did not dispute, at least in her articles, women's belonging to the private sphere, but instead linked this role with a commentary on women's education.[40] Her endorsement of the gender binary coincides with nineteenth-century Latin American feminist discourses that spoke to the importance of women for the republic while nonetheless assuming traditional gender roles. As Davies, Owen, and Brewster show, early nineteenth-century Latin American writers had already upheld sexual difference as a way to justify women's ambivalent inclusion within foundational political projects.[41] Further, as Nancy Cott's study of the English context makes evident, the appropriation of sexual difference was key in the development of nineteenth-century feminist movements because it allowed women to argue "for equal opportunities in education and employment, and for equal rights in property, law and political representation, while also maintaining that woman would bring special benefits to public life by virtue of their particular interests and capacities."[42] In other words, the binary had been useful for women to justify their role in society and subsequently push for the improvement of women's rights. However, as Méndez's life demonstrates, she was constantly pushing the boundaries and asserting herself in the public sphere in a way that she did not necessarily endorse for other women. She understands herself as an exceptional woman but does not envision her public endeavors as a model for Mexican women in general.

From an ecofeminist perspective, Val Plumwood categorizes this movement as "the feminism of uncritical equality or Artemisian feminism" and defines it as "another term for 'liberal feminism', based on the story of the goddess Artemis, which 'has tended to accept the basic structures of existing political and economic institutions, pressing hardest on the need to make them accessible to women."[43] Méndez's feminism is indeed an "Artemisian feminism" that naturalized racial, class, and social hierarchies in its attempt to open new spaces for women to participate in public and political affairs. Méndez's *Revista* is an important example of "Artemisian feminism" because it is intertwined with a push toward the formation of a strong economic transnational community in Latin America. The discourses of the magazine make evident that for an elite Latin American woman, stereotypes of gender difference were a gateway to promote an extractivist pedagogy that could counter the global hierarchies of market trade and development. The feminization of nature becomes an important part of Méndez's extractivist pedagogy. Further, *La Revista*'s use of female images to represent the abundant and fertile Latin American lands become an illustration of how Méndez approached feminism not only as a tool to participate in the public debates of her time but also as an strategy to inscribe Latin America in global notions of civilization.

Ultimately, Méndez's extractive pedagogy promoted two main principles: first, educating elite Latin Americans in the importance of appropriating the profits and particularities of the extractivist model; and second, the formation of a transnational economic community that could take advantage of the potential wealth of the Latin American lands. Both goals were supported by a reification of a set of hierarchical dichotomies that supported stereotypical conceptions of womanhood, the nonhuman, nature, and the body. The magazine's pedagogical initiative was rooted in an understanding of education, more specifically extractivist education, as the only means to alter the uneven distribution of wealth between the global North and South and inscribe Latin America as more than a peripheral contributor. The overarching message of *La Revista*, beyond its language differences, relied on western ideas of progress and civilization, supported the cultural primacy of the United States, legitimized an extractive relation between humans and nature, and reified notions of gender difference. The extractivist pedagogy was dependent on English and Spanish sections alike. *La Revista*'s English sections educated the English-speaking audience in the wealth and potential development of Latin America while enabling elite Latin Americans to understand the kind of relation Anglo-Americans established with the land. The Spanish section reinforced those examples with descriptions about California's success and by creating a homogeneous representation of Latin America as a fertile, abundant space just patiently waiting to be exploited. In the end, both languages worked as a means to promote the same extractivist relation, either in the form of foreign exploitation or of governmental initiatives.

La Revista's extractivist pedagogy demonstrates how the hierarchical dichotomies of reason and nature also inform Méndez's feminism. The use of female tropes to represent the fertility and abundance of the land becomes a signifier of her own understanding of the movement. For Méndez, feminism was an important path to inscribe Latin America in the trends of civilization. In her understanding, the movement was not trying to question sexual differences, instead the reification of those differences along with a gendered representation of the land was crucial for Latin Americans to understand the economic and political importance of appropriating the extractivist model. Therefore, from *La Revista* to Méndez's feminist articles, and even in "La curva," we can identify the same intention. Namely that Latin Americans needed to understand how important exploiting the land was in order to address the disparities between North and South, alter the global hierarchies of civilization and market trade, and help establish a successful transnational Latin American community. Just as in the cautionary tale of "La curva," in *La Revista*'s discourse egalitarian relationships are promised as a result of, first, the strengthening of a Latin American community and, second, an understanding of the need to appropriate the dynamics and profits of the extractivist model.

One year and seven months after *La Revista* was founded, by September 1896, Mendez's shareholders in *La Revista*, the professor Harold Howard and the translator Thomas Savage, deprived the Mexican writer of her share in the periodical. Mendez hadn't signed a formal contract and therefore was obliged to resign the direction of a periodical that was already renowned and profitable.[44] The swindle that Méndez depicted in "La curva" could be seen as an iteration of her own experience. *La Revista* continued to publish until December 1896, with four issues under the direction of Thomas Savage. Méndez lost control of her enterprise and returned to Mexico to focus solely on her role as an educator. However, the notions that she had endorsed through the magazine, especially her extractivist pedagogy, traversed her subsequent short stories, miscellaneous articles, education reports, manuals, and travel writings.

NOTES

1. Méndez, "La curva," 374.
2. Méndez, "La curva," 374.
3. Bazant, "La educación moderna."
4. Laura Méndez attended conferences on public education in Saint Louis, Missouri, and Germany and wrote extensive reports for the Porfirio Díaz government. The reports were recently published in the book *Laura Méndez de Cuenca: Su herencia cultural*, volume 3.

Méndez used the institutional reports in subsequent articles written for a general audience and published in Mexican periodicals such as *La mujer mexicana* and *El imparcial*.

5. In her article "Las necesidades de México: México necesita educación" (Mexico's needs: Mexico needs education), Méndez says "En la época presente la educación es un ave rara que como el pájaro azul del cuento, no se llega a ver sino tras de penalidades y trabajos inacabables" (In this present time education is a rare fowl that as the blue bird from the story, is only seen after hardship and endless burdens; 170).

6. Bazant, *Laura Méndez de Cuenca*.

7. Mílada Bazant's book *Laura Méndez de Cuenca: Mexican Feminist, 1853–1928* uncovers Méndez's different roles in late-nineteenth-century Mexican society. Bazant's subsequent three-volume project *Laura Méndez de Cuenca: Su herencia cultural* gathers Mendez's poetry, translations, short stories, chronicles, miscellaneous articles, feminist articles, travel chronicles, and educational texts and pairs them with critical studies by Pablo Mora, Leticia Romero Chumachero, Ana Rosa Domenella, Luzelena Gutiérrez de Velasco, Roberto Sánchez Sánchez, and Lilia Granillo.

8. Méndez's identity as an elite Mexican was intertwined with an understanding of how Anglo-Americans saw and approached not only Latin American lands but also Latin American people. *La Revista*'s double discourse made this awareness evident by rhetorically differentiating articles written in English from those written in Spanish. Méndez's double rhetoric could be related to what W. E. B. Du Bois defines as "double consciousness" when talking about the African American experience in the United States. See Du Bois, *The Souls of Black Folk*. However, Méndez's *Revista* was not necessarily providing a contesting representation of Latin America. Instead, the magazine appropriated Anglo-American discourses about Latin America as a way to promote an extractive relation between elite Latin Americans and their land.

9. Eduardo Gudynas provides a detailed definition of the concept in the article "Extracciones, extractivismos y extrahecciones."

10. Pratt, *Imperial Eyes*, 124.

11. Mary Louise Pratt defines "contact zones" as "the space of imperial encounters, the space in which peoples geographically and historically separated come into contact with each other and establish ongoing relations, usually involving conditions of coercion, radical inequality, and intractable conflict." Pratt, *Imperial Eyes*, 8. Méndez's *Revista* was produced in late nineteenth-century San Francisco, a place that witnessed the interactions of white Americans traveling to establish new business posts, Indigenous communities being displaced of their territory and California residents that identify more with Mexico's dynamics and customs.

12. Milada Bazant talks about Mendez's years in San Francisco in her book *Laura Méndez de Cuenca*.

13. Méndez, "Epistolario," 618.

14. English articles focused mainly on calculating potential profits and detailing the kind of businesses that were not only needed but also a guaranteed success. For instance, an article from July 1895 titled "Signs of the Times" says: "Our correspondents and exchanges, the facts gathered from the United States consular reports and bulletins, and interviews with visitors from those countries confirm the bright outlook and speak of nothing but abundant crops, high prices and general prosperity. . . . The contracts on one plantation alone amount to $1,000,000." Méndez and Howard, "Signs of the Times."

15. From the first issue to the fourth, the magazine included a total of sixteen articles about California.

16. See Méndez and Howard, "California despierta"; Méndez and Schleiden, "Una valiosa biblioteca"; Méndez and Howard, "Digno de aplauso."

17. Méndez's preoccupation with the problem of neocolonialism becomes apparent with the publication of "La curva"; within the magazine the topic is not addressed in the same terms. At the beginning of "La curva" the narrator emphasizes how Silverio and his family had changed nationality without realizing it: "los ascendientes de Silverio [que] . . ., habían cambiado de nacionalidad allá por el año de 47, así como el borrico cambia de ronzal: sin darse cuenta" (Silverio's ancestors . . . had changed nationalities around the year of 47 exactly as the donkey changes halter, without realizing it; 370).

18. Méndez and Schleiden, "California y sus elementos," 5.

19. Méndez and Schleiden, "California y sus elementos," 5.

20. Méndez and Schleiden, "California y sus elementos," 5.

21. Méndez and Schleiden, "California y sus elementos," 5.

22. See Anderson, *Imagined Communities*.

23. Méndez and Schleiden, "Nuestra posición," 3.

24. Méndez and Howard, "Córdoba," 5.

25. Méndez and Howard, "Estado de México," 5.

26. Ericka Beckman explores the relation between products and discourses of nationality in her book *Capital Fictions: The Literature of Latin America's Export Age*.

27. Diego von Vacano traces the complexities of the creation and later erasure and synthesis of race in the Latin American context in his book *The Color of Citizenship*.

28. The magazine's promotion of a Latin American community resembles the homogenizing ideologies created by creole elites to promote the formation of nationalist discourses during the Age of Revolutions. Marixa Lasso outlines the particularities of this discourses, emphasizing how creoles erased racial differences in an attempt to homogenize Latin American communities and promote unity in her book *Myths of Harmony: Race and Republicanism during the Age of Revolutions*.

29. Martí said: "Nations that remain strangers must rush to know one another, like soldiers about to go into battle together. Those who once shook their fists at each other like jealous brothers quarreling over who has the bigger house or who owns a plot of land must now grip each other so tightly that their two hands become one." Martí, "Our America." Marti's idea of knowing each other and understanding America as a means to achieve a good government is echoed in the purpose of *La Revista*. However, Marti's call for new and independent political forms is not present in Méndez's magazine.

30. Val Plumwood explains the relation between the canonical western understanding of reason as the high realm and the ecofeminist movements in her book *Feminism and the Mastery of Nature*.

31. Méndez and Schleiden, "California y sus elementos," 5.

32. Méndez and Schleiden, "California y sus elementos," 5.

33. Méndez and Schleiden, "California y sus elementos," 5.

34. Merchant, *The Death of Nature*.

35. See Romero Chumacero, "Un impulso de solidadridad," 195; and Bazant, *Laura Méndez de Cuenca*, 84.

36. Romero Chumacero, "Un impulso de solidadridad," 191.

37. Méndez, "El decantado feminismo," 218.
38. Romero Chumacero, "Un impulso de solidaridad," 197.
39. Méndez, "El decantado feminismo," 217.
40. In her book *Laura Méndez de Cuenca*, Milada Bazant shows how throughout her life, Méndez challenged and broke with the traditional female role of the "angel of the house."
41. Davies, Owen, and Brewster, *South American Independence*, 4.
42. Cott, *The Grounding of Modern Feminism*, 20.
43. See Plumwood, "Gender, Eco-Feminism," 16.
44. The details of this polemic are described in a letter that the Mexican sends to her friend the journalist Enrique Olavarría y Ferrari in August 1896: "La causa de mi tardanza en contestar ésta tiene que referirse a un golpe que mi apreciable socio me dio en la chapa del alma, quedándose con el periódico y sus pertenencias todas, por haber yo confiado en su lealtad y descuidado el contrato de sociedad" (The cause of my tardiness to reply its linked to a blow that my dear partner gave me in the soul when he kept the entirety of the periodical because I trusted his loyalty and did not procure a partnership contract). See Méndez, *Epistolario*, 622.

BIBLIOGRAPHY

Anderson, Benedict R. *Imagined Communities: Reflections on the Origin and Spread of Nationalism*, rev. ed. Verso, 2006.

Bazant, Mílada. "La educación moderna según el lente crítico de una mujer ilustrada." In *Laura Méndez de Cuenca: Su herencia cultural*. Siglo XXI Editores, 2011.

Bazant, Mílada. *Laura Méndez de Cuenca: Mexican Feminist, 1853–1928*, 2nd ed. Translated by Mary Kay Vaughan. University of Arizona Press, 2018.

Beckman, Ericka. *Capital Fictions: The Literature of Latin America's Export Age*. University of Minnesota Press, 2013.

Cott, Nancy F. *The Grounding of Modern Feminism*, rev. ed. Yale University Press, 1989.

Davies, Catherine, Hilary Owen, and Claire Brewster. *South American Independence: Gender, Politics, Text*. Reprint, Liverpool University Press, 2011.

Du Bois, W. E. B. *The Souls of Black Folk*. Myers Education Press, 2018.

Ferrús Antón, Beatriz. "Y ves palidecer tu luz hermosa: La poesía de Laura Méndez de Cuenca." In *Casa en que nunca he sigo extraña: Las poetas hispanoamericanas: Identidades, feminismos, poéticas (Siglos XIX–XXI)*. Peter Lang, 2017.

Gudynas, Eduardo. "Extracciones, extractivismos y extrahecciones: Un marco conceptual sobre la apropiación de recursos naturales." In *Observatorio del desarrollo*, no 18. Centro Latino Americano de Ecología Social, 2013.

Lasso, Marixa. *Myths of Harmony: Race and Republicanism during the Age of Revolution, Colombia, 1795–1831*. University of Pittsburgh Press, 2007.

Martí, José. "Our America." University of Pennsylvania Libraries, January 1, 1891. https://writing.upenn.edu/library/Marti_Jose_Our-America.html.

Méndez, Laura. "El decantado feminismo." In Milada, *Laura Mendez de Cuenca*.

Méndez, Laura. "Epistolario." In Milada, *Laura Mendez de Cuenca*, vol. 2.

Méndez, Laura. "La curva." In Milada, *Laura Mendez de Cuenca*, vol. 2.

Méndez, Laura, and Harold Howard. "California despierta." April 1895. Hemeroteca Nacional, Biblioteca Nacional de México.

Méndez, Laura, and Harold Howard. "Cordoba." March 1895. Hemeroteca Nacional, Biblioteca Nacional de México.

Méndez, Laura, and Harold Howard. "Digno de aplauso." May 1895. Hemeroteca Nacional, Biblioteca Nacional de México.

Méndez, Laura, and Harold Howard. "Estado de México." May 1895. Hemeroteca Nacional, Biblioteca Nacional de México.

Méndez, Laura, and Harold Howard. "Signs of the Times." July 1895. Hemeroteca Nacional, Biblioteca Nacional de México.

Méndez, Laura, and José Schleiden. "California y sus elementos de vida." February 1895. Hemeroteca Nacional, Biblioteca Nacional de México.

Méndez, Laura, and José Schleiden. "Cauca." March 1895. Hemeroteca Nacional, Biblioteca Nacional de México.

Méndez, Laura, and José Schleiden. "Colombia." March 1895. Hemeroteca Nacional, Biblioteca Nacional de México.

Méndez, Laura, and José Schleiden. "Nuestra posición en la prensa." February 1895. Hemeroteca Nacional, Biblioteca Nacional de México.

Méndez, Laura, and José Schleiden. "Una valiosa biblioteca." February 1895. Hemeroteca Nacional, Biblioteca Nacional de México.

Merchant, Carolyn. *The Death of Nature: Women, Ecology, and the Scientific Revolution*, illustrated edition. HarperOne, 1990.

Milada, Bazant, ed. *Laura Méndez de Cuenca: Su herencia cultural.* 3 volumes. Siglo XXI Editores, 2011.

Mora, Pablo. "Laura Méndez de Cuenca: Pasión y destino en la poesía mexicana." In Milada, *Laura Méndez de Cuenca.*

Plumwood, Val. *Feminism and the Mastery of Nature.* Routledge, 1994.

Plumwood, Val. "Gender, Eco-Feminism and the Environment." In *Controversies in Environmental Sociology*, edited by Robert White. Cambridge University Press, 2004.

Pratt, Mary Louise. *Imperial Eyes: Travel Writing and Transculturation*, 2nd ed. Routledge, 2007.

Romero Chumacero, Leticia. "Un impulso de solidadridad: El feminismo de Laura Méndez." In Bazant, *Laura Méndez de Cuenca.*

Romero Chumacero, Leticia. *Una historia de zozobra y descontento: La recepción de las primeras escritoras profesionales en México.* Editorial Gedisa, 2018.

von Vacano, Diego. *The Color of Citizenship: Race, Modernity and Latin American/Hispanic Political Thought.* Oxford University Press, 2012.

Warren, Karen J. *Ecofeminist Philosophy: A Western Perspective on What It Is and Why It Matters.* Rowman & Littlefield, 2000.

CHAPTER 10

Memories of a Darwinian

Anarchism and Animality in the *Literary* Crónicas *of Rafael Barrett*

JENNIFER L. FRENCH

I. Lo que son los yerbales: Animalizing Humans

One of the more unsettling aspects of Rafael Barrett's writing—including his exposé of labor and environmental abuses in the yerba mate industry, *Lo que son los yerbales* (1908)—is his use of animal imagery to represent human beings, especially marginalized or subaltern Paraguayan workers and their children.[1] In one typical example, the Hispano-Paraguayan anarchist and author (1876–1910) evokes two kinds of "animals" involved in the harvesting of yerba: the indebted worker is a "poor, frightened beast," while the overseer becomes a "ferocious beast." He also writes:

> Así trabaja hozando en el bosque sus galerías de topo, tendidas de picada a picada, agujeros en fondo de saco por donde busca y trae la yerba. Desgaja, carga y acarrea el ramaje al fogón. Se arrastra penosamente bajo el peso que le abruma. A eso se reduce la estúpida faena del yerbal, a la de una acémila que hocicara ante su sendero de retorno.[2]

This is how the laborer roots through his moles' galleries in the forest, opened up by pick-ax strokes, holes in the bottom of the cloth sack he uses to carry the yerba. He tears off, carries and hauls the branches to the bonfire. He drags himself painfully under the overwhelming weight. This is what the mindless labor of the yerbales comes down to, the work of a mule snuffling along its circular route.

And, of course, there is the opening segment of *Lo que son los yerbales*, called "Degeneración," in which Barrett writes, "Escudriñad bajo la selva: descubriréis una criatura agobiada en que se van borrando los rasgos de su especie. Aquello ya no es un hombre: es un peón yerbatero" (Peer beneath the jungle: you will discover an overwhelmed creature in whom the features of his species are being erased. That is no longer a man: it is a yerba worker; 1:320). Here is one more example, the supply is nearly limitless: in "Lo que he visto," from the posthumous collection *El dolor paraguayo*, Barrett describes the misery of the Paraguayan countryside. His rhetoric passes fitfully across the human/nonhuman divide from one order of being to another, pausing to focus on the physical features of undernourished women and children. The mothers are represented as exhausted flesh, wounded even in their reproductive organs, while starving children become "niños esqueletos, de vientre monstruoso, los niños arrugados, que no ríen y no lloran, las larvas del silencio" (skeletal children, with monstrous bellies, wrinkled children who neither laugh nor cry, larvae of silence; 1:249).

Such passages are discomfiting because they literally dehumanize the rural Paraguayans Barrett was otherwise defending. What Zakkiyah Iman Jackson calls "racialized animalization" and "bestializing social logics" have a long history in the Atlantic world, where Christian theology's hierarchical ordering of species and insistence on the uniqueness of the human soul have frequently been mobilized to vilify non-white/non-European peoples by association with supposedly "lower" orders of being and to legitimate exploitation and violence against both.[3] In the mid-nineteenth century, Domingo F. Sarmiento's notion of a conflict between civilization and barbarism invoked this principle and updated it for a modern, secularizing Río de la Plata region, where it became the organizing concept behind the Argentine liberals' campaigns against Juan Manuel de Rosas (1839–1852) as well as the War of the Triple Alliance against Paraguay (1864—1870). A few years after Rosas's defeat, the Guarani-speaking Paraguayans, led by the autocratic Francisco Solano López, became the epitome of savagery, backwardness, irrationality, and animality, a threat to the stability of the region that could only be resolved by a process of near-extermination.[4]

As Jason W. Moore writes, "capitalism does not 'produce' nature in a linear fashion, but is an evolving whole that joins the accumulation of capital, the pursuit of power, and the co-production of nature."[5] Risky as it is, Barrett's rhetorical strategy may be said to evoke "nature" as co-produced in the aftermath of a genocidal war and the postwar Paraguayan governments' decision to liquidate the country's remaining *tierras fiscales*, or publicly owned agricultural lands: displaced campesinos, including survivors of the war and their descendants, migrated toward the yerba groves, also recently privatized, and became ensnared in a legalized system of debt-slavery. In Barrett's rendering, Paraguayan bodies—no longer required to defend the State militarily—have become what Moore refers to as "Cheap Nature," the undervalued resources of labor, food, energy, and raw materials, typically located at the frontier of capitalist expansion, on which accumulation depends. As Moore writes, "Cheap Nature is produced when the interlocking agencies of capital, science, and empire—blunt categories, yes—succeed in releasing new sources of free or low-cost human and extra-human natures for capital."[6] Some readers may disagree, but I continue to see flickers of ambivalence in passages like the ones quoted here; nevertheless, I also recognize given how frequently he lambastes racist discourses of the time—that Barrett's jarring figurative language is intended to convey not inherent animality so much as the ways that war and privatization (also known as "capital, science, and empire") actively shift the frontier between the human and the subhuman, or in the terms suggested by biopolitical theory, between the life that is to be protected and preserved and life that is rendered expendable.[7]

Moore's theory builds on the argument, forcefully advanced by the Peruvian sociologist Aníbal Quijano, among others, that European colonization of the Americas in the sixteenth and seventeenth centuries, acting in the service of an emergent capitalism, at once hardened the long-standing divide between "Man" and "Nature" within Western thought and attempted to eradicate Amerindian ontologies that recognize radically different forms of being, including the agentiality of multi-species assemblages.[8] Quijano writes, "The fact that Western Europeans will imagine themselves to be the culmination of a civilizing trajectory from a *state of nature* leads them also to think of themselves as the moderns of humanity and its history, that is, as the new and, at the same time, most advanced of the species."[9] Evidence of other ways of understanding the world—or in less Eurocentric terms, knowledge of worlds that do not conform to European assumptions about reality—was taken as confirmation of the "backwardness" of non-European societies, their irrationality and failure to evolve beyond the original "state of nature."[10]

Of course, such knowledge persists in many forms throughout the Americas. In the Paraguayan context, we find fascinating traces of it in the archive of the Triple Alliance War, the massive national mobilization that sparked, among other developments, important shifts in the use of the Guarani language. For the first time since the days of the Jesuit missions, Paraguayan *letrados* set about determining how to represent Guarani with roman letters so as to produce newspapers that soldiers could read aloud to one another in the military encampments. Today the articles are considered of secondary interest to the ingenious *xilograbados* or woodcut prints created by Paraguayan soldier-artists to illustrate the pages of these *periódicos de trincheras*. As art historians Ticio Escobar and Osvaldo Salerno comment, the scarcity of academy-trained artists in wartime Paraguay meant that the symbols, forms, and sensibility of the country's popular culture are all on display in the images.[11] The front page of every number of *Cabichuí* features an image of stinging wasps—called *cabichuí* in the Guarani language—swarming around a single human figure, suggesting a multitude of small-but-mighty Paraguayan soldiers harassing an Afro-Brazilian. Inside the paper, images of animal transformation abound. One typical print shows an orange tree being climbed by three figures in soldier's clothing; their smooth round heads have enormous beaks and their feet are talons. The caption reads, "A estos bandidos se coge con las naranjas, como a las peces con la carnada" (These bandits can be caught with oranges, like fish can be caught with bits of meat).[12] The Allied leaders are constantly subject to animalization: the Baron de Porto-Alegre is a swimming capibara, complete with epaulets and kepi; Bartolomé Mitre is a harp-strumming dog in a dress, lectured by a well-dressed peacock identified as the Marqués de Caxias; the Argentine General Juan Andrés Gelly y Obes is a tall skinny *oveja*, or sheep.

The critic and creative artist Josefina Plá comments that the zoomorphic hybrids of *Cabichuí*, like the art produced in the seventeenth-century missions, suggest the ongoing significance of the Guarani culture in the collective unconscious of Paraguay's popular classes.[13] According to the Mybá cosmology, for example, when the gods destroyed by flood the first land, *Yvy tenonde*, people who did not reach the state of divine grace known as *aguyje* were transformed into plant and animal species. "Later," writes anthropologist María Victoria Cebolla Badie, "with the creation of *Yvy pau*,

A estos bandidos se coge con las naranjas, como á los peces con la carnada.

FIGURE 10.1. "These bandits can be caught with oranges, like fish can be caught with bits of meat." Anonymous. Woodcut print from *Cabichuí*, June 6, 1867. Courtesy of The Museo de Barro, Asunción.

Porto Alegre, (tirándose de cabeza)--Fufs....Qué diabos dos paraguayes..., tenho muito calor !

FIGURE 10.2. "To hell with the Paraguayans . . . it's so hot!" Anonymous. Woodcut print from *Cabichuí*, October 3, 1867. Courtesy of The Museo de Barro, Asunción.

FIGURE 10.3. "Mitre: 'Those suckers the Brazilian royal family will get eaten up when the Conquest of Paraguay fails.' Caxias: 'The Argentine president will be the dog that eternally sniffs the world's butt, having failed to get a clue from the front.'" Anonymous. Woodcut print from *Cabichuí*, December 23, 1867. Courtesy of the Museo de Barro, Asunción.

Mitre—La realeza brasilera será el pavo de la boda en el fracaso de la conquista del Paraguay.
Caxias—El General Presidente Argentino será el perro que olerá eternamente al mundo por detras, ya que no lo ha entendido por delante.

the new land, they populated the jungles, fields and rivers, retaining attributes from when they were human beings."[14]

As Eduardo Viveiros de Castro writes, one of the key distinctions between European and Amazonian ontologies is that the West sees human beings as having risen out of their previously animal condition, whereas the myths of many Amerindian societies describe how nonhuman animals came to be deprived of their human condition: "Humans are those who continue as they have always been: animals are ex-humans, not humans ex-animals."[15] We know that Barrett listened carefully to the stories of his Paraguayan interlocutors, whom he cites repeatedly in *El dolor paraguayo* and elsewhere as the source of stories of animal-becomings and other transformations. Barrett describes Paraguayan campesinos as the "inheritors" of Guarani thought, endowed with a language tied to the natural world so inextricably that it "varies from place to place, forming dialects within a dialect that is one of many in central South America."[16] In "La poesía de las piedras"—one of the *crónicas* republished posthumously in *El dolor paraguayo*—he writes that the campesinos understand "la tristeza de las piedras" (the sadness of stones; 1:225) as the desolation of human beings who were transfigured in an act of divine retribution. "Las bestias-oráculos," another of the *crónicas* collected in *El dolor paraguayo*, seems at first glance a simple enumeration of superstitions associated with various nonhuman creatures, but here too Barrett incorporates evidence of the transformational nature of the Guarani world. Barrett repeats the story of a neglectful son who was transformed into a bird called a *caraú* (1:231) and describes the dangers of the fox, who "ha hecho en el bosque alianzas con espíritus malignos" (has made alliances with malignant spirits in the forest; 1:232).[17] And then there is the mysterious and threatening creature known in urban legend as "El Pombero," of whom Barrett writes, "Si no fuera por su mirada inteligente, se le creería un animal, el animal más parecido al hombre" (If not for his intelligent gaze, he would be considered an animal, the animal that most closely resembles man).[18]

Considered in this context, the transformations Barrett represents in *Lo que son los yerbales* may suggest the influence of autochthonous practices of worlding; we might even say that he identified a point of convergence between Guarani ontology and the European notion of degeneration ("or devolution"), which was often used in medical discourse to represent the trans-generational effects of social ills such as alcoholism and sexual abuse.[19] In *Lo que son los yerbales* this convergence yields a polysemic trope that Barrett uses to expose, if only partially, the untold horrors being perpetrated on the frontier of capitalist expansion, to illustrate a story, that is, about capitalism and its effects, *seen and unseen*, on Paraguay's multiple worlds.[20] If there remains any doubt that Barrett realized there was more to the pillaging of Paraguay's natural yerba groves than meets the (modern/Western) eye, we may consider as well the final paragraph of "Herborizando," a vignette first published in March 1908 and later included in *El dolor paraguayo*. In these lines, Barrett describes the curative powers associated with various plants known to Paraguayan herbalists, culminating in a discussion of *yerba* as "a sovereign of ancient lineage," that has "heard everything, divined everything: terrible confidences, ruined hopes, somber oaths . . . a thousand blended voices of the immense past."[21] As Rodrigo N. Villalba Rojas comments, Barrett is a collector of Indigenous and Paraguayan myths who understands their role in the campesino struggle as what Villalba Rojas calls "a mode of emancipation" (un modo de emancipación).[22]

Anthropologist Mario Blaser, reiterating a point he attributes to his Indigenous interlocutors, stresses the significance of storytelling in the co-constitution of ontologies and the worlds they convey: "stories are not only or not mainly denotative (referring to something 'out there'), neither are they fallacious renderings of real practices; rather, they partake in the variably successful performance of that which they narrate." What this means, according to Blaser, is that "the stories being told cannot be fully grasped without reference to their world-making effects," and thus if stories may be said to "'world' worlds," then some stories world better worlds than others, and "some 'world' worlds we do not want to live in or with."[23] In what follows, I turn from Barrett's ethnographic and political writings to his more explicitly scientific works, specifically the *crónicas* compiled posthumously in a collection titled *Mirando vivir* and the essay "Filosofía del altruismo." They show us a Barrett who avidly followed scientific developments in Europe and presented them to audiences in the Río de la Plata through his columns in the daily press. While there is much more to say about this topic than can be covered in a single chapter, with Blaser's comments in mind, I am particularly interested in the stories that Barrett tells about and with European science. For contemporary anthropologists like Blaser, the New Materialisms and closely related fields like Science and Technology Studies (STS) and Actor-Network Theory provide a useful frame of reference for understanding that "reality is always in the making through the dynamic relations of heterogeneous assemblages involving more-than-humans."[24]

Barrett was working within the confines of European science as it existed in the very early years of the twentieth century, well before many of the developments in theoretical physics, molecular biology, and other fields to which the New Materialisms respond. What is more, Barrett was not an academic scientist or philosopher: his entire corpus, which runs to more than 1,500 pages, was written for the daily press and produced under increasingly precarious conditions.[25] Contradictions and inconsistencies abound; there are few attempts at systematization in Barrett's writing. There are even fewer footnotes, which makes pinpointing his sources extremely difficult. Nevertheless, I will suggest that Barrett engages with the science of his day in ways that push back on what many of his contemporaries saw as its political implications while also—and perhaps less obviously—creating cognitive frameworks better suited to accommodating Paraguay's multiple worlds.

II. Mirando vivir: *Science and Politics*

My reading of Barrett builds on recent scholarship by Francisco Corral Sánchez-Cabezudo, Ana María Vara, and Leandro Delgado. Their work represents a significant shift from the earlier studies of Barrett's writings to which we are all nevertheless indebted. Long condemned to the margins of literary studies by a trifecta of seemingly irremediable "faults"—his expatriate and transnational status, his unequivocally radical politics, and the journalistic nature of his writing—Barrett's reputation in the Americas was maintained during the latter half of the twentieth century by a cadre of exceptionally distinguished readers, all of them conspicuously situated on the political left: Miguel Ángel Fernández, Augusto Roa Bastos, Ángel Rama, and David Viñas. The compilation of Barrett's writings edited by Fernández and published by Biblioteca Ayacucho in 1978 as *El dolor paraguayo* established enduring interpretative paradigms in terms of the selection of texts, which favors Barrett's acerbic commentary on politics in the Río de la Plata, as well as the content and tenor of Roa Bastos' introductory essay. Framing Paraguayan history in the language of Developmentalism and Cold War politics, Roa Bastos underscores Barrett's personal integrity and his "anarquismo humanista y moralizador" (humanist and morally minded anarchism).[26] The emphasis on Barrett's "humanism," while strategically unassailable in terms of the distance it creates between the intellectual tradition of anarchist thought and the stereotypical *dinamitero*, nevertheless makes readers less likely to notice Barrett's engagement with animality and nonhuman nature more broadly. Similarly Viñas, introducing selections from Barrett's work in *Anarquistas en América Latina* (1983), describes Barrett's remarkable erudition as manifest in *El dolor paraguayo*: the list of references spans from Pope Leo XIII to the Boer War, Ibsen, Tolstoy, and the Dreyfus Affair, but does not extend beyond literature and politics as conventionally construed.[27]

More recently, scholars have expanded consideration of Barrett's work beyond *El dolor paraguayo* and begun to engage with his scientific background and his fascination with the more-than-human. In this regard the contributions of the Spanish philosopher Francisco Corral have been decisive: prior to editing Barrett's two-volume *Obras completas* (Tantín, 2010), he published the monograph *El pensamiento cautivo de Rafael Barrett: Crisis de fin del siglo, juventud del 98 y anarquismo* (1994), which traces Barrett's evolving political positions and the influences on his thinking, particularly the writings of Nietzsche, Piotr Kropotkin, and Henri Bergson. Corral is a key source for an excellent ecocritical reading of Barrett by the Argentine critic Ana María Vara, who identifies Barrett as a powerful subterranean influence in the development of *Ríoplatense* literature, and in particular as the source of a new mode of writing about capitalism's exploitation of humans and nonhuman nature that would surface later on in writers including Horacio Quiroga, Roa Bastos, and Eduardo Galeano. "Barrett 'socializa' la naturaleza," she writes, "encuentra la sociedad en la naturaleza" (Barrett "socializes" nature . . . he finds society in nature).[28] My reading of Barrett draws on the scholarship of Corral and Vara as well as Leandro Delgado's outstanding work as editor of Barrett's *Crónicas de la naturaleza*, a collection of short texts in which anarchism, ethnography, and science are all intertwined in the tight space of the literary *crónica*. Delgado's prologue emphasizes Barrett's debt to fellow anarchists like Kropotkin and Elisée Reclus, but also his standing as their intellectual equal, his intermittent suspension of anthropocentric thinking, and his compassion for a more-than-human nature "that also suffers the incomprehensible actions of man."[29]

Given Barrett's eventual ousting from upper-class Spanish society, much has been made of his mother's connection to the Peninsular aristocracy, but it was his father—an English mathematician and accountant—who oversaw the early education of Rafael Barrett and his brother Fernando.[30] His education and training were remarkably cosmopolitan: Barrett spoke English, French and Spanish fluently; he attended secondary school in Paris before entering the Escuela de Ingeniería in Madrid.[31] Like the rest of the Spanish *Generación del 98*, Barrett had experienced at first hand the intense ideological battles touched off by the translation of Darwin's major works, which became a flashpoint in the struggle between conservative Catholics and the liberalizing Krausian school.[32] Corral maintains that Barrett recognized early on that Darwin's theory of evolution was a positive, liberating counter to the Catholic Church's emphasis on eternal verities and unquestionable dogmas, but also that he initially understood the promised liberation in the purely individualistic terms inspired by his reading of Nietzsche and Herbert Spencer. After emigrating to the Río de la Plata and becoming immersed in the social, economic, and political realities of the region— and those of Paraguay above all—Barrett would change his position dramatically: from 1906 on he vehemently rejected the social Darwinism that dominated mainstream intellectual circles in Europe and the Americas, with its vision of an "individualistic, liberal and stateless society" as "the culmination of human evolution."[33] He embraced instead Kropotkin's principle of mutual aid, which aims to excise from Darwinian thought the influence of social pessimists like Malthus, Huxley, and Spencer, emphasizing that the primary mechanism for change within species is not competition over scarce resources but rather a propensity for association.[34]

Throughout his years in the Río de la Plata (1904–1910), Barrett published *crónicas* and other writings that draw on his scientific background to interpret developments in Europe and the United States on behalf of local readers.[35] Their titles give a sense of the range of topics covered: "El amor" (April 1905), "Óptica" (December 1905), "Amatistas" (December 1906), "Mosquitos" (August 1907), "Dactiloscopia" (October 1907), "La ciencia," (September 1908), "Tarifa médica" (February 1907), "La moral y la ciencia" (August 1909), "Vivisecciones," (November 1909), "Halley" (December 1909), "Naturología" (December 1909), "El mito naturista" (February 1910), "Perros" (May 1910), "Hormigas" (March 1910), "La evolución de los mundos" (April 1910), "La ciencia y el crimen" (July 1910), and "Vacuna" (July 1910). In a few cases Barrett cites sources like the London *Times*; more commonly he does not. *Crónicas* on closely related topics will appear in different papers in quick succession, and Barrett intermittently refers to his own recent writings, seemingly regardless of where they were published.[36] According to Corral these pieces, while scattered among different papers published in Asunción and Montevideo, constituted a regular column dedicated to science, politics, and society under the wonderfully suggestive title *Mirando vivir.*[37] The transdisciplinary nature of Barrett's thinking is inherent in the texts: he is not simply disseminating scientific information so much as interpreting the news according to his own shifting subjective and political criteria. And, in fact, when Barrett openly espouses anarchism, he emphasizes the importance of scientific principles as one of the key tenets in his politics, writing in "Mi anarquismo" (November 1906) that human society should be regulated by the same principles that govern science: "Ningún sabio, por ilustre que sea, presentará hoy su autoridad como un argumento; ninguno pretenderá imponer sus ideas por el terror" (No scientist, however illustrious, sets forth his own authority as an argument; none would try to impose his ideas through terror; 2:801). Barrett's *crónicas* frequently call out self-serving interpreters of scientific fact—be they Catholic clergy, secular positivists, or neoromantic naturopaths—at the same time that he uses his own aestheticized treatment of evolutionary theory to redirect their philosophical imaginings along lines more in keeping with his own priorities.

Darwin himself is only intermittently invoked in Barrett's writing, be it because of the ideological association with social Darwinism, or because of the vilification of non-European peoples in Darwin's books—particularly the Indigenous inhabitants of Tierra del Fuego whom he encountered during the 1831 voyage of the HMS *Beagle*—or for some other reason. As Cannon Schmitt writes, Darwin's recollections of the Fuegians were hardly tangential to the development of the theory of evolution: on the contrary, "prosecuting the argument for human descent from other animals requires that Darwin find in Fuegians and other 'savages' representatives of an otherwise empty space in the continuum of living forms, a missing link; accordingly, he does just that."[38] In subsequent decades Darwin's theory, above all the notion of the survival of the fittest, would be invoked to justify colonialism and genocide in many parts of the world, including Paraguay and the southern territories claimed by Argentina and Chile. That said, like Kropotkin and Reclus, Barrett seems less inclined to contradict Darwin himself than to differentiate between what he considers legitimate and illegitimate interpretations of Darwin's ideas. On October 26, 1906, for example, Barrett delivered a public lecture in which he corrected popular misconceptions regarding Darwinian thought, saying it had been subjected to distortions "sólo favorable a los cristianos que descienden modestamente de Dios Padre Todopoderoso" (favorable only to the Christians who modestly descended from God the All-Mighty Father; 2:498). Exactly three months later he takes a different tone in "Los colmillos de la raza blanca" (Fangs of the white race), a caustic commentary on European wars of conquest that includes the lines, "Humanos que no sois blancos, creed en el misionero y su frasco de alcohol, en el traficante y en su látigo de negrero, creed en Jesucristo y en Darwin, porque son lo mismo" (Humans who are not white, believe in the missionary and his alcohol, in the middleman and his slave-catcher's whip, believe in Jesus Christ and in Darwin, because they are the same; 2:52). One might assume, on the basis of that line, that Barrett was renouncing Darwinism altogether, but that is not the case.[39]

Barrett deals with Darwinian questions of human agency, futurity and planetary life in a second set of dialectically entwined *crónicas* from the same period, "La tierra" and "El retorno a la tierra." In the first piece, published in an Asunción paper called *Los Sucesos* on August 23, 1906, Barrett reports a series of earthquakes that recently occurred in California and Chile; in response he declares his commitment to human freedom and criticizes religious leaders who are taking advantage of the situation to reassert their dubious authority. He praises a powerful, heroic humanity that is capable of piercing mountains and tearing the earth open to make it fertile; in Barrett's view the cataclysm on the western coasts is merely a temporary setback to this mighty (and mightily phallic) humankind. Three weeks later "El retorno a la tierra" appears in *El Cívico*.[40] The tenor is entirely different:

La enseñanza profunda del siglo XIX es la de nuestra identidad con la naturaleza. Hemos descubierto que los fenómenos físicos obedecen a leyes, es decir, a fórmulas intelectuales. La realidad se encaja en los moldes de la razón como la llave en su cerradura. Pero no es sólo nuestra inteligencia la que, sobre la enorme y luminosa superficie del universo, se mezcla con su propia sangre, parecidamente a esos anchos árboles que hunden su follaje en los ríos, besando la sombra que tiembla sin cesar bajo las aguas; nuestra sensibilidad, nuestra carne perecedera y dolorosa se ha revelado hermana de la humilde carne de las bestias. La arquitectura de nuestros cuerpos se ha revelado la misma: el mismo nuestro oscuro origen y el juego de nuestros instintos; la misma, quizá, nuestra destinación misteriosa. (1:676)

The profound lesson of the nineteenth century is that of our identity with nature. We have discovered that physical phenomena obey laws, that is, intellectual formulas. Reality fits within the molds of reason like a key within a lock. But it is not only our intelligence that mixes with its own blood upon the luminous surface of the universe, like those wide trees that sink their foliage into a river, kissing the shadow that trembles unendingly beneath the waters; our sensitivity, our mortal and aching flesh has been revealed as sister to the humble flesh of beasts. The architecture of our bodies reveals the same: the same our obscure origin and the play of our instincts; the same, perhaps, our mysterious destination.

The sense of confidence in humanity's intellectual ability conveyed in "La tierra" is reiterated here, but tempered: reason does not set humanity apart from the rest of nature so much as confirm our deep enmeshment with other forms of being. Barrett offers a fascinatingly complex analogy comparing the relationship between human intelligence and the phenomenological universe to the relationship between a tree and the waters of a river into which its leaves dip; then he abruptly shifts into a more straightforward style that recalls the presentation of evidence in *The Descent of Man*. Darwin's use of comparative anatomy and comparative behavior is recapitulated in Barrett's references to "nuestra sensibilidad," "nuestra carne perecedera y dolorosa," "la arquitectura de nuestros cuerpos," and "el juego de nuestros instintos." (For reference, Darwin writes, "It is notorious that man is constructed on the same general type or model with other mammals, all the bones in his skeleton can be compared with corresponding bones in a monkey, bat, or seal."[41] And a page later, "Many kinds of monkeys have a strong taste for tea, coffee, and other spiritous liquors: they will also, as I have myself seen, smoke tobacco with pleasure."[42]) However, Barrett is clearly more interested in transcending the Darwinian framework than he is in reproducing it point by point; to that end, he emphasizes an intuition of kinship that extends across wildly different species and forms of being: sunlight, river, and plants, human and nonhuman animals. All are tied together in his original analogy, which becomes increasingly literal as the paragraph unfolds. Like Darwin, Barrett suggests that all of these beings are becoming something different than what they are, with an important distinction: in Darwin's theory, transformation tends toward ever greater differentiation among species, "by and for the good of each being," whereas Barrett, true to the anarchist spirit, invokes instead a process of collective becoming.[43] Human and nonhuman life forms advance toward a mysterious destination that is at once "ours" and "the same" as that of everything else.

Viveiros de Castro writes that "Western popular evolutionism is very anthropocentric, but not particularly anthropomorphic," whereas "'primitive animism' may be characterized as anthropomorphic, but it is definitely not anthropocentric: if sundry other beings besides humans are 'human,' then we humans are not a special lot."[44] What he means is that as Darwin's theory entered mainstream discourse in the Atlantic world, it was adjusted— "distorted," Barrett would say—so as to preserve the special status that Western societies traditionally attribute to the human. In the modern Western ontology, Man is believed to have "surpassed" the other animals in his "development," and the other animals are of interest only inasmuch as they anticipate Man and may be exploited by him; Amazonian perspectivism instead situates human and nonhuman animals on a horizontal plane but "privileges" the human insofar as it attributes human-like motives and perceptions to nonhuman animals. In the passage quoted above and in what follows, Barrett's response to Darwin's theory of evolution is both anthropomorphic and anthropodecentric. As "Retorno a la tierra" concludes, we see that the title refers not only to a follow-up or continuation of the earlier *crónica*, "La tierra," but also to a belief that Darwinism—"the profound lesson of the nineteenth century"—sparks (a desire for) more intimate and equitable relations with the earth itself:

> No somos ya hijos de los dioses. No está ya nuestra grandeza en el pasado, sino en el futuro. No es de arriba y de lejos de donde nos viene la vida, sino que nos envuelve, nos abraza, nos penetra. Semejantes a las plantas, sentimos las partes elevadas de nuestro ser besadas y agitadas por el viento libre, al tiempo que nuestras raíces, largas y tenaces, nos atan cada vez mejor a las tinieblas fecundas. Y he aquí por qué amamos la tierra más sólidamente, más lúcidamente, más humanamente. (I: 676)

> We are no longer children of the gods. Our greatness lies not in the past, but rather in the future. Life does not come to us not from up above and far away, but rather it enfolds us, embraces and penetrates us. Like plants, we feel the elevated parts of our being kissed and stirred by the free breeze, at the same time as our roots, long and tenacious, tie us ever more firmly to the fecund darkness. And for this we must love the earth more solidly, more lucidly, more humanly.

Much could be said about the ways that this beautiful passage evokes premodern European modes of thought, beginning with the etymology of *humano*, which comes to us from the Latin *hŭmus* for earth—as in *inhumar*, *exhumar*, and the modern word *humus*—and stems from the medieval universities' distinction between the secular, earthly branches of knowledge and the realm of theology. As Joan Corominas comments in his *Diccionario etimológico de la lengua castellana, humano* is related "only more distantly to *homo*, or man."[45] Barrett's meaning here is consistent with that etymology, since he admonishes readers not to lament our collective loss of status but rather to revel in and relish our real place in the universe, which is not separate from other life-forms but rather fully enmeshed in them.[46] We are also reminded of what Michel Foucault refers to as "the old analogy of plant to animal (the vegetable is an animal living head down, its mouth—or roots— buried in the earth)."[47] For my purposes, however, it is important to point out that Barrett's writing resonates with some non-Western ontologies as well as with premodern Europe's: both Pachamama and Mother Earth find expression in the turn away from the aggressive, phallic tropes of "La tierra" toward the apperception of a "fecund darkness" that requires respect and recognition as the source of earthly life, a "fecund darkness" upon which we humans are ineradicably dependent. In contrast to "La tierra," in which a heroic humanity operated collectively through the use of massive machines capable of violently tearing and manipulating the earth, "El retorno a la tierra" finds us assuming the form and condition of plants: earth-bound for our survival but capable of feeling the embrace of vital energies that "enfold" and "penetrate" us from the "fecund darkness" below. Barrett is not only tinkering with scale and sexual imagery: crucially, his imagining of the human is derived from an anthropomorphic imagining of vegetal life. If we are like plants, he suggests, plants are also like us: capable of feeling, sensing, and *enjoying* physiological exchanges with the sources of nutrients and energies that sustain and ultimately transform them.

At the same time, Barrett's insistence that we love the earth *humanamente*—which could be rendered in English as humanly or humanely— resonates with his writings about the extremely inequitable distribution of land in post-war Paraguay. I am thinking in particular of the *crónicas* that respond to campesino demands for agricultural land by representing their relationship with the earth in terms of mutual affection and care. For example, in "La tierra," a speech delivered to a workers collective in the early months of 1908, Barrett exhorts his audience to "emancipate" the land and defend it from parasitic landowners who know nothing of agricultural labor and thus have no claim to more than the narrow burial plot that will ultimately enclose their bodies. He refers here to the land as "sustentadora de cuanto alienta, fuente de inmortalidad" (sustainer of all that breathes, fountain of immortality; 2:521). I am reminded also of Barrett's 1907 review of *Cuadros históricos y descriptivos del Paraguay*, Cecilio Báez's analysis of the Triple Alliance War. Barrett concludes his review by emphasizing the need to restore the old, affective ties between Paraguayan campesinos and the land. He represents the postwar liquidation of *tierras fiscales* as tantamount to the massive mortality of the war itself and describes the Paraguayan campesinos as "orphans" cut off from their loving parent. He insists on the need to re-establish the "vital thread" of rural agricultural and artisanal traditions and "reanudar en fin las amorosas, íntimas, capitales relaciones con el terruño" (re-tie, at last, the loving, intimate, capital relations with the land; 2:254). Surely Delgado correct in saying that Barrett privileges the kind of knowledge "that is reached by collective spirits when they recognize, in contact with nature, that the souls of men come from the relation with trees and grasses, with 'superstitions,' with the souls of stones and things without distinguishing among species [or] races."[48]

III. Deep Time in "Filosofía del altruismo" and "Sobre el Atlántico"

In reality, the writings I have been calling "Darwinian" also demonstrate the increasing influence of Henri Bergson—himself an important interpreter of Darwin—whose philosophical reconceptualization of time as duration would lead in directions Barrett evidently found exceptionally provocative. Corral speculates that Barrett may have been reading Bergson as early as his Paris years, when Barrett likely studied mathematics under the famous polymath Henri Poincaré; whatever that history may turn out to be, Barrett's 1907 essay "Filosofía del altruismo" bears the unmistakable imprint of Bergson's *Creative Evolution*, which was published in French the previous year. The encounter with *Creative Evolution* might even be said to have prompted Barrett to attempt a rare systematization of his thinking, inasmuch as "Filosofía del altruismo" offers at once a sustained critique of positivism and a theory of human subjectivity that draws on Barrett's diverse intellectual commitments: anarchist politics, Darwinism, philosophical vitalism, and the emergent field of psychoanalysis. While a full examination of this fascinating text lies outside the scope of the present chapter, the brief analysis I offer here is intended to illuminate Barrett's engagement with Darwin and Bergson and its significance for his critique of capitalism.

Barrett introduces the essay by telling readers not to expect academic philosophizing so much as an articulation, in what he calls the imperfect medium of human language, of an inner trajectory such as "todo ser vivo" (every living being) senses within itself, and perhaps even "todo cristal y todo átomo" (every crystal and every atom; 2:527). Thus defying conventional Western ontology in favor of a more animistic understanding of reality, he distinguishes between "lo verdadero" (truth), the object of science that changes from century to century, and "lo real," the object of wisdom that concerns everyone: "De lo verdadero nos servimos; de lo real, vivimos, o por mejor decir, lo real es lo que vive" (Truth is something we make use of, whereas we live off the real, or better said, the real is what lives; 2:529). Corral notes that Barrett's use of "lo real," if philosophically imprecise, is nevertheless indebted to Bergson's famous concept of *élan vital*, the vital impetus that survives from generation to generation, providing both continuity across time and the most fundamental form of difference, that between the living and the dead.[49] With that trans-individual flow in mind, Barrett wonders where the *yo* is located, where it begins and where it ends (2:530). He feels within himself the presence of an impersonal force hurtling toward an undefined future:

> un total incoherente que necesita mudar de actitud y esperar lo que no ha sucedido todavía . . . algo irresistible que se opone a la estéril repetición del pasado, y que ansía romper las barreras del egoísmo para realizar su obra imposible.

> an incoherent whole that needs to change its attitude and wait for what has not yet happened . . . something irresistible that opposes the sterile repetition of the past, and that longs to break the boundaries of egotism to realize its impossible task (531).

As we have seen in other writings, here too Barrett explicitly rejects competitive social Darwinism; he advocates instead an anarchistic vitalism that attends to matter, bodies, and the real. Bergson's philosophy of change is thus harnessed to the boundary-breaking politics of the radical left. The bounded individual of Western philosophy is vehemently rejected; what Barrett offers instead is a trans-specific vitality that mysteriously acts on matter, including the body it inhabits, and that is most alive when it delivers itself to the flow of other vital forces in the universe.

"¿Se puede hacer una filosofía de metáforas?" (Can a philosophy be made of metaphors?; 2:529). In another apparent nod to Bergson, Barrett asks this interesting question to introduce the figurative language he will use to describe the mysterious interiority of the subject: the invisible world, the secret world we carry within us.[50] We might ask him in response, what is metaphor and what is real? Or, where does metaphor end and the real begin? Here as in his literary *crónicas*, Barrett is a dazzlingly virtuosic writer, using alliteration, apostrophe, ironic turns and shifts of pace as well as analogies and elaborate extended metaphors.[51] While we know that Barrett was committed to the artful and even highly aestheticized use of language, there is arguably something more literal—something more "real"—happening when Barrett refers to his own vital force as "sister" to that of a flowering plant, alike in their unconscious and impersonal desire to perpetuate their existence into a distant and unknowable future: "Mi energía directiva, hermana de la humilde energía celular que convierte los jugos de la tierra en pétalos perfumados, pasará a través de vuestras leyes como el viento cargado de gérmenes a través de una tela de araña" (My executive energy, sister of the humble cellular energy that transforms the juices of the earth into perfumed petals, will pass through your laws like the wind carrying seeds through a spider's web; 2:534).

For Barrett, the real is the deep continuity across life-forms separated by time and space, the persistence of past into present that represents the shared evolutionary history of all living things. Barrett's awareness of that continuity and the history behind it intermittently gives rise to the feeling of trans-specific kinship that the feminist philosopher Elizabeth Grosz— herself a meticulous reader of Darwin and Bergson—refers to as "becoming undone," the stimulating and strangely comforting awareness that comes from recognizing "one's own personal smallness, but also one's fundamental connection to almost everything."[52]

Bergson himself calls this feeling "solidarity," a term that gets us even closer to the meaning of Barrett's Darwinian tropes. In the particular case of "Filosofía del altruismo," he shifts the emphasis so that the drive to political action—anarchist boundary-breaking—becomes all but indistinguishable from the irrepressible generative processes of more-than-human forms of life.[53] Barrett seizes on Bergson's insistence that evolution be understood as a truly open-ended becoming—a distinction that Bergson famously explained in terms of the contrast between "virtual multiplicity" and "actual multiplicity," whereby the latter erroneously reduces the possible to a mirror image of what actually exists—and he makes it a manifesto for the creative, emancipatory power that life itself bestows upon the living.[54] "Hermanos, vivís," he writes, "somos lo nuevo; estamos fuera de la ley" (Brothers and sisters, you are alive, we are the new; we are outside the law; 2:535). In Barrett's rendering, *élan vital* and the real are both used interchangeably with the concept of altruism, which he describes as itself "fuera de las leyes" (outside the laws; 2:535). Following anarchist scientists of the time, he delves into Spencer's infamous argument that survival requires species to adapt to their environment; Barrett produces from natural history counterexamples of creatures that ensure their individual and collective longevity by actively altering their environments instead: "¿Qué hace la humanidad," he asks, "sino humanizar el universo?" (What does humanity do, if not humanize the universe?; 2:535).[55] Adaptation to society's written and unwritten laws is certain death, and life demands a collaborative, collective rebellion that is nothing less than the unleashing of the vital impulse, *élan vital*, or as Barrett himself puts it, "la marcha moral de las energías creadoras" (the moral progress of creative energies; 2:525).

Barrett evokes the evolutionary intuition of "becoming undone" in "Filosofía de altruismo" and in *crónica* after *crónica*, so frequently that it may be considered a rhetorical signature. For example, in "De estética," published in *El Diario* on August 29, 1905, he directly echoes *The Descent of Man*, in which birds occupy four lengthy chapters supporting the theory of sexual selection whereby individuals of certain species are held to choose their mates on the basis of aesthetic qualities, i.e., attractiveness.[56] Barrett comments, "Nos es lícito inducir que el arte en el hombre, como en el amblyornis, es un fenómeno sexual" (We may deduce that in man as in the bower bird, art is a sexual phenomenon; 2:479). In "El río invisible" (*El Diario*, November 22, 1907), he writes, "Todos llevamos en nosotros una historia tan antigua y venerable como la creación misma" (We all bear within us a story as ancient and venerable as creation itself; 1:108). In "Indumentaria" (*La Razón*, March 15, 1910), "Hemos superpuesto tantas naturalezas en nuestro ser, que ya no sabemos cuál es la primera ni cuál es la más importante" (We have superimposed so many natures on our being, that we no longer know which is the first nor which the most important; 2:89).

Among these Darwinian *crónicas* the most affecting is almost certainly "Sobre el Atlántico," (On the Atlantic; 2:428–29), which was published in Montevideo's *La Razón* on October 4, 1910. One of the last works Barrett would publish during his brief life, it represents an evolutionary meditation on the trans-oceanic voyage he undertook seeking an experimental treatment for the tuberculosis that had afflicted him for nearly four years. (The treatment, prescribed in Paris by a Dr. Quintón, involved injections

Mirando vivir

Sobre el Atlántico

«C'étaient les eaux, et les eaux, et les eaux»
(Jammes)

Las aguas parecen sin fin, como si no hubiese ya tierras, y nuestro mundo fuera una inmensa gota, una sola y redonda lágrima azul cayendo en el éter. ¡Oh, este azul! Es un azul obscuro, denso, translúcido, un azul de zafiro en cuyo seno, bajo las alas de la noche, despiertan fulgores de fósforo. ¡Dónde la espuma sería más blanca que sobre el azul á veces laminado y bruñido como un metal, á veces laqueado de negro, el azul atlántico que me llena la vista y el alma? Espuma rodante, sonora, cabellera de nieve salvaje, penacho que se alza y se arroja y se levanta nuevamente y se encabrita en cada cresta del innumerable y paralelo ejército de olas. Espuma, — surtidor, torrente, cascada — que en lo cóncavo de la onda teje anchos exágonos irregulares cuyas cintas tiemblan como sobre una piel, ó que adelgaza sus filamentos lívidos en un encaje de sutileza infinitesimal, ó se desvanece en verde bruma submarina, ó se curva en gasa que se deshace al viento, ó se retuerce en largas volutas de humo líquido, ó finge á los oblicuos rayos del sol la red de púrpura que inyectara el ojo enorme de un monstruo... Espuma blanca sobre el mar azul, emulsión hirviente de agua y aire... Sí; aire, agua, nada más; lo que cede y se desliza y huye, y por lo mismo rodea y devora y disuelve. Agua y aire, lo que carece de cohesión y de forma, y por lo mismo revela su inflexible geometría en el arco fatal del horizonte...

Aguas del mar, estremecidas y desnudas, sangre purísima del universo, linfa madre, plasma sagrado del cual llevamos todos, para poder vivir, una provisión en las venas! Tu sal se seca en mis labios, y saboreo tu sublime amargura. Acaso á una legua bajo la quilla del buque yacen las ruinas de un continente que recuerdan los hombres — y acaso cien otras bajo ellas —, pero en tus entrañas surgen continuamente las Venus primordiales; seres blandos y errabundos, tentáculos ciegos, larvas glaucas, pulpa ancestral que se ha vuelto transparente y flota invisible, bosques sumergidos, infinitas lianas de un ámbar sin flor, y también el semillero de la fauna microscópica, polen oceánico que en vastas estelas arde bajo el firmamento de los trópicos. Y quizá, en una hora tibia, oh mar venerable! engendras aún, como en las épocas geológicas, el misterio de los misterios, las células matrices de la vida virgen...

¿Aún?... Nada hay ilimitado ni eterno. El mar envejece. Su aliento se pierde en los espacios siderales. Su agua, cristalina limpieza entregada á los cielos, le es devuelta avaramente por los ríos, turbia y sucia, cargada de todos los despojos y secreciones y deyecciones de la tierra. Y con el transcurso de los tiempos el mar se torna más aere, más espeso, más bajo, más árido. Nosotros los siempre más ágiles, los usurpadores del destino, corremos hoy sobre las aguas, cortándolas al doble tajar de nuestras hélices, porque supimos aprisionar el fuego, y el fuego, como nos anunció Esquilo, es el maestro que nos lo ha enseñado todo, todo!— hasta fabricar lo álgido y helar el

aire. ¿Qué importa que se apaguen los astros, si se encienden otros en nuestros cerebros? Y todavía mañana, cuando el mar haya cuajado en un témpano único sus sueños estériles, volarán nuestras máquinas sobre él, dejando en las tinieblas un rastro de chispas.

R. B.

FIGURES 10.4 AND 10.5. Rafael Barrett, "Sobre el Atlántico" (On the Atlantic). Image courtesy of Fundación Rafael Barrett.

of seawater.)[57] Miraculously, the original manuscript in Barrett's own impeccable hand has survived. He begins by describing the vision that presents itself on deck on a dark night, imagining an act of reverse creation in the immensity of waters that stretch out around the boat in every direction "como si no hubiese ya tierras" (as if lands no longer existed; 1:428). The tone is sedate, but the speaker is nonetheless mesmerized by what he calls the "inflexible geometría" (inflexible geometry; 1:428) of the forms. Then he writes:

¡Aguas del mar, estremecidas y desnudas, sangre purísima del Universo, linfa madre, plasma sagrado del cual llevamos todos, para poder vivir, una provisión en las venas! Tu sal se seca en mis labios, y saboreo tu sublime amargura. Acaso a una legua bajo la quilla del buque yacen las ruinas de un continente que recuerdan los hombres—y acaso cien otras bajo ellas— pero en tus entrañas surgen continuamente las Venus primordiales: seres blandos y errabundos, tentáculos ciegos, larvas glaucas, pulpa ancestral que se ha vuelto transparente y flota invisible, bosques sumergidos, infinitas lianas de un ámbar sin flor, y también el semillero de la fauna microscópica, polen oceánico que en vastas estelas arde bajo el firmamento de los trópicos. Y quizás, en una hora tibia, ¡oh mar venerable! engendras aún, como en las épocas geológicas, el misterio de los misterios, las células matrices de la vida virgen. (1:428–29)

Waters of the sea, shaken and naked, purest blood of the Universe, mother-lymph, sacred plasma from which we all carry, to enable our life, a provision in our veins! Your salt dries on my lips and I taste your sublime bitterness. Perhaps a league beneath the keel of the ship there lie the ruins of a continent remembered by men—and perhaps a hundred more beneath them—but in your innards primordial Venuses continually arise: soft and wandering beings, blind tentacles, glaucous larvae, ancestral flesh that has become transparent and floats invisibly, submerged forests, infinite vines of an unblooming amber, and also the seedbed of microscopic fauna, oceanic pollen that on vast stelae burns beneath the firmament of the tropics. And perhaps, at a tepid hour, oh venerable sea! you yet engender, as in geological ages, the mystery of mysteries, the womb-cells of virgin life.

Barrett's language is stirring, his imaginings more vivid and original than ever. ("Ningún abatimiento de espíritu," Barrett notes in his medical diary the previous week: "la capacidad intelectual parecería haber aumentado" (My spirit is not discouraged; my intellectual abilities seem to have expanded; 2:678). As Donna Haraway would put it, drawing out the etymology of "tentacle" from the Latin *tentaculum* (feeler) and *tentare* (to feel), Barrett is making kin with the "tentacular ones," allowing himself and the reader to perceive our common belonging in a trans-specific community of life. Riffing ingeniously on Darwin's identification of the remotest origins of earthly life in a creature resembling "the larvae of our existing marine Ascidians," Barrett conjures these tiny beings, ur-mothers of the deep, not as the anthropomorphic sea goddess of Botticelli and the classical Mediterranean tradition, but as a multitude of real creatures enmeshed in a real lifeworld, part of a complex deep-sea assemblage that is increasingly threatened by capitalist modernity.

In the third and final paragraph the tone becomes more somber as Barrett announces that the sea has grown acrid, heavy, and "dry," its breath "lost" in space and its waters increasingly contaminated by the secretions and detritus of human settlements on the land. It is clear from Barrett's writing that the responsible party is not all of humanity, but rather those who have access to modern Western technology, especially internal combustion engines like the ones that power his ship:

> Nosotros, los cada vez más ágiles, los usurpadores del destino, corremos hoy sobre las aguas, cortándolas al doble tajar de nuestras hélices, porque supimos aprisionar el fuego, y el fuego, como nos anunció Esquilo, es el maestro que nos lo ha enseñado todo, ¡todo!, hasta fabricar lo álgido y helar el aire. ¿Qué importa que se apaguen los astros, si se encienden otros en nuestros cerebros? Y todavía mañana, cuando el mar haya cuajado en un témpano único sus sueños estériles, volarán nuestras máquinas sobre él, dejando en las tinieblas un rastro de chispas. (1:429)

> We, the ever-more-agile, usurpers of destiny, run today across the waters, cutting them with the double blades of our propellers, because we learned to capture fire, and fire, as Aeschylus announced, is the master that has taught us everything, everything! Even how to fabricate coldness and freeze the air. What does it matter if the stars are extinguished, if new ones are lighted in our brains? And even tomorrow, when the sea's sterile dreams have congealed in a single ice floe, our machines will fly across it, leaving a trail of sparks in the darkness.

In contrast to "Filosofía del altruismo," "El retorno a la tierra," and other *crónicas*, "Sobre el Atlántico" has no optimistic futurology in store, no thrilling evocation of the evolution of life-forms beyond the human.[58] Nor does Barrett indulge here in the comforting idea of human mortality as an enfolding of the self into the ceaseless cycles of nature—on the contrary, the text invokes what for twenty-first century readers will be an eerily familiar sense that natural processes of decline and renewal have been profoundly disturbed.[59] His articulation of a collective subject of disorder also has a contemporary ring, as when Barrett refers to "nosotros, los cada vez más ágiles, los usurpadores del destino," emphasizing the illegitimate harnessing of world-altering technologies on behalf of individual ambition. Infinite becomings have been stopped in their tracks: the sea is dying and humanity—or more precisely, the narrow subset of humanity able to take advantage of the combustion engine—has killed it. Barrett is not generally technophobic, but the passage demonstrates a profound ambivalence about the unnatural (dis)order created and perpetuated by global capitalism. It's hard not to picture Elon Musk blasting off in his private spaceship when Barrett writes, "¿Qué importa que se apaguen los astros, si se encienden otros en nuestros cerebros?"

Augusto Roa Barrett observes that Barrett "mostró cómo era posible producir textos de valores intrínsecos y autónomos; que no se proponían la simple transcripción de la realidad visible sino la mostración y revelación de la realidad invisible en la virtualidad de sus múltiples significaciones" (showed how it was possible to produce texts of intrinsic and autonomous value; texts that did not propose a simple transcription of visible reality but that offered instead the demonstration and revelation of invisible reality in the virtuality of its multiple significations).[60] It is a remarkably timely description of a writer still commonly associated with social realism. To read Barrett through Darwin and Bergson is to appreciate more fully his idiosyncrasy as an anarchist scientist writing literary *crónicas* on the frontier of capitalist expansion; it is also to tease out and identify the hidden registers of Barrett's language that Roa Bastos calls its *virtuality*, or in Deleuze's own parlance, "the cloud of virtual images" that surrounds "every actual."[61] Barrett's persistent evocations of the evolutionary past, present, and future of human and other life forms suggest the almost-inconceivable potentialities of unseen worlds and, ultimately, their fragility. In that regard, his writings echo back and amplify the concerns of our own time. As we have seen intermittently throughout this chapter, Barrett's writings resonate with the

Deleuzian currents that run through contemporary scholarship in the environmental (post)humanities. This is neither coincidental nor anachronistic, of course: Barrett is a Deleuzian thinker *avant la lettre* because Barrett and Deleuze share fundamental intellectual commitments and an imaginary in which Darwin and Bergson loom very large.[62] Through his increasingly anarchistic interpretation of their works, Barrett develops a philosophical system, however loosely articulated, based on change and multiplicity as well as an imagistic repertoire that he draws on to tell stories about capitalism and its effects, seen and unseen, on multiple worlds in Paraguay and far beyond.

NOTES

1. I would like to thank Miguel Ángel Fernández, Guido Rodríguez Alcalá, Ximena Briceño, Héctor Hoyos, Giuseppina Forte, Mari Rodríguez Binnie, and Gene Bell-Villada for their comments on earlier versions of this chapter.
2. Barrett, *Obras Completas 1*, 319. Hereafter references to Barrett's work are given parenthetically.
3. Jackson, *Feminist Studies* 39, 674. See also Jackson, *Becoming Human*. On the history of "racialized animalization" and related tropes in Latin American contexts, see Ximena Briceño, "Animal," especially pages 91–96. See also Jennifer French and Gisela Heffes' introduction to *The Latin American Ecocultural Reader*.
4. Capdevila, *Una guerra total*, 65–66.
5. Moore, *Capitalism in the Web of Life*, 19.
6. Moore, *Capitalism in the Web of Life*, 53.
7. Leandro Delgado has also addressed the question of animalization in *El dolor paraguayo* and other of Barrett's works. See Delgado, "Prólogo," in Barrett, *Crónicas de la naturaleza*, 14–16. The classic formulation of laissez-faire liberalism as the governing modality particular to biopolitical power is Michel Foucault's *The Birth of Biopolitics*. On biopolitics and animality in Latin American cultural production, see Giorgi, *Formas comunes*.
8. See, for example, Quijano, "Coloniality of Power"; and Mignolo, "Commentary."
9. Quijano, "Coloniality of Power," 191, emphasis added.
10. Since the late 1990s, anthropologists have engaged in vigorous debates about what has come to be known as the "ontological turn" in the discipline. For a succinct analysis of the debates and their stakes see Mario Blaser, "Ontology and Indigeneity." I take the notion of "multiple realities or worlds" from Blaser, 52 and forward.
11. Escobar and Salerno, "El grabado de *Cabichuí*."
12. *Cabichuí*, June 6, 1867, 2.
13. Plá, "El grabado."
14. Cebolla Badie, "Cosmología y naturaleza," 142–43.
15. Viveiros de Castro, "Cosmological Deixis," 472.
16. The quoted phrases are from Barrett, "La poesía de las piedras," 225, and "Guaraní," 223.
17. Barrett, "Las bestias-oráculos," 229–32.
18. Barrett, "El Pombero," 238. See Delgado, "Prólogo," 15–16.
19. See Ellenberger, *The Discovery of the Unconscious*, 280–81.

20. My use of "multiple worlds" here is informed by Mario Blaser's analysis of debates within anthropology's ontological turn; see Blaser, "Ontology and Indigeneity," 49–58.

21. See French, "Yerba."

22. Villalba Peña, "Rosicrán y su poema," 41.

23. Blaser, "Ontology and Indigeneity," 54.

24. Blaser, "Ontology and Indigeneity," 54.

25. As Francisco Corral Sánchez-Cabezudo observes, in one of his final letters Barrett "expressed a desire to exchange the brief and rapid form of the article for the density and repose of a book, lamenting that the practical hardships obliged him to dedicate himself exclusively to journalism." Corral Sánchez-Cabezudo, "Introducción," *Obras completas*, vol. 1, 31.

26. French and Heffes, "Developmentalism," 211–14; Roa Bastos, *El dolor paraguayo*, xxvii.

27. Viñas, *Anarquistas*, 180.

28. Vara, *Sangre que se nos va*, 66.

29. Delgado, *Crónicas de la naturaleza*, 13.

30. On the causes, context and consequences of Barrett's "desclasamiento" see Corral, *El pensamiento cautivo*, 6–19.

31. Muñoz, *El pensamiento vivo*, 15–21, and Corral, *El pensamiento cautivo*, 4. Although the social scandal that prompted his emigration to the Río de la Plata prevented him from completing a degree, Barrett was considered a brilliant and accomplished mathematician. He published at least two mathematical papers in Madrid in the 1890s. After emigrating to the Río de la Plata in late 1903, Barrett resolved a problem formulated by the French savant Henri Poincaré and received a congratulatory letter from Poincaré himself. Barrett was intermittently employed as a math tutor and was involved in the founding of the Unión Matemática Argentina, a precursor of the engineering school at the University of Buenos Aires. While it was journalism that initially took Barrett to Paraguay—covering the 1904 revolution for *El Correo Español*—when the victorious Liberales incorporated Barrett into their new government it was in the Office of Statistics and subsequently in Engineering and Railways. He would teach courses in advanced math and offer lectures at the Instituto Paraguayo, an important cultural and educational institution in Asunción, and was published in its journal (1905). When Barrett's increasingly trenchant criticism of the Liberales and his support of the working classes set him at odds with the Asunción establishment, he would once again make use of his mathematical skills and training, finding work as a surveyor in the surrounding countryside.

32. Corral Sánchez-Cabezudo, *El pensamiento cautivo*, 4.

33. Hale, "Political Ideas." Also see Corral, *El pensamiento cautivo*, 110–16; Glick, Ruiz, and Puig-Samper, eds., *El darwinismo*.

34. Ferretti, "Evolução and Revolução."

35. According to the Paraguayan historian Claudio Fuentes Armadans, Barrett likely obtained his information from sources that were widely available to the Paraguayan elite of the time, including Argentine magazines like *Caras y Caretas* and *PBT*, which reached Asunción and the interior regularly. Claudio Fuentes Armadans, personal communication, January 17, 2024.

36. Viriato Díaz Pérez, extensively cited by Carlos R. Centurión in his classic work of cultural history *Historia de las letras paraguayas*, indicates that among the newspapers to which Barrett contributed, only *El Diario* and *Los Sucesos* remunerated him for his work. Centurión, 318.

37. Corral, "Presentación de la colección," 19.

38. Schmitt, *Darwin and the Memory*, 27.

39. Ana María Vara offers a nuanced interpretation of "Los colmillos de la raza blanca" in *Sangre que se nos va*, 126–30.

40. *Los Sucesos* and *El Cívico* were both published in Asunción and owned by prominent ParaguayanLiberales.

41. Darwin, *The Descent of Man*, 784.

42. Darwin, *The Descent of Man*, 785.

43. Darwin, *On the Origin of Species*, 760.

44. Viveiros de Castro, "Cosmological Deixis," 484–85.

45. Corominas, *Breve diccionario*, 327.

46. See Donna Haraway and Rusten Hogness' riff on the link between humanities and humus in Haraway's *Staying with the Trouble*, 32.

47. Foucault, *The Order of Things*, 21.

48. Delgado, Prólogo, xxviii.

49. Corral, *El pensamiento cautivo*, 133.

50. On Bergson's use of metaphor as an element of his philosophical method see L. Lawlor, "Henri Bergson." *Stanford Encyclopedia of Philosophy*, 2004. Online at https://plato.stanford.edu/entries/bergson/#CreaEvol.

51. On Barrett's use of literary language, see Delgado, xvii and French, "Yerba," 89–102.

52. Grosz and Stirner, "All Too Human," 19. I am grateful to Mary Louise Pratt for introducing me to Grosz's work.

53. "An organism such as a higher vertebrate is the most individuated of all organisms; yet, if we take into account that it is only the development of an ovum forming part of the body of its mother and of a spermatozoon belonging to the body of its father, that the egg (i.e., the ovum fertilized) is a connecting link between the two progenitors since it is common to their two substances, we shall realize that every individual organism, even that of a man, is merely a bud that has sprouted on the combined body of both its parents. Where, then, does the vital principle of the individual begin or end? Gradually we shall be carried further and further back, up to the individual's remotest ancestors: we shall find him solidary with each of them, solidary with that little mass of protoplasmic jelly which is probably at the root of the genealogical tree of life. Being, to a certain extent, one with this primitive ancestor, he is also solidary with all that descends from the ancestor in divergent directions. In this sense each individual may be said to remain united with the finality of living beings by invisible bonds." Bergson, *Creative Evolution*, 49–50.

54. See Keith Ansell Pearson and John Ó Maoilearca, Introduction to *Key Writings*.

55. Corral, *Pensamiento cautivo*, 115.

56. Grosz discusses the theory of sexual selection extensively in *Becoming Undone*, 115–201.

57. Muñoz, *El pensamiento vivo de Barrett*, 43.

58. I take the concept of "futurology" from Mary Louise Pratt. See *Planetary Longings*, especially 121 and forward.

59. On Barrett's comforting invocation of natural cycles, see Delgado's reading of "En la estancia" in his prologue to *Crónicas de la naturaleza*, xxi–xxii.

60. Roa Bastos, "Rafael Barrett," xxix.

61. Deleuze, "The Actual and the Virtual," 148.

62. On Deleuze's relation to Bergson and Darwin, see Keith Ansell Pearson's book, *Germinal Life: The Difference and Repetition of Deleuze*, and Elizabeth Grosz's writings including *Becoming Undone.*

BIBLIOGRAPHY

Barrett, Rafael. *Obras completas*, vol. 1–2, edited by Francisco Corral. Tantín, 2010.

Bergson, Henri. *Creative Evolution*. Translated by Arthur Mitchell. Modern Library, 1944.

Bergson, Henri. *Essential Writings*. Bloomsbuy, 2002.

Blaser, Mario. "Ontology and Indigeneity: On the Political Ontology of Heterogeneous Assemblages." *Cultural Geographies* 21, no. 1 (2014): 49–58.

Briceño, Ximena. "Animal." In *Handbook of Latin American Environmental Aesthetics*, edited by Jens Andermann, Gabriel Giorgi, and Victoria Saramago. De Gruyter, 2023.

Capdevila, Luc. *Una guerra total: Paraguay, 1864–1870. Ensayo de historia del tiempo presente.* Translated by Ana Couchonnal. Editorial sb and CEADUC, 2010.

Cebolla Badie, María Victoria. *Cosmología y naturaleza mbya-guaraní*. PhD diss. Universitat de Barcelona, 2013.

Centurión, Carlos R. *Historia de las letras paraguayas*, vol. 2. Editorial Asunción, 1948.

Corominas, Joan. *Breve diccionario etimológico de la lengua castellana*. Gredos, 1987.

Corral Sánchez-Cabezudo, Francisco. "Introducción." In *Obras completas*, vol. 1, by Rafael Barrett; edited by Francisco Corral-Cabezudo, 25–31. Tantín, 2010.

Corral Sánchez-Cabezudo, Francisco. "Introducción." In *Obras completas*, vol. 2, by Rafael Barrett; edited by Francisco Corral-Cabezudo, 21–27. Tantín, 2010.

Corral Sánchez-Cabezudo, Francisco. *El pensamiento cautivo de Rafael Barrett: Crisis de fin del siglo, juventud del 98 y anarquismo*. Siglo XXI, 1994.

Corral Sánchez-Cabezudo, Francisco. "El pensamiento de Rafael Barrett, un 'joven del 98' en el Río de la Plata." *Revista de Hispanismo Filosófico*, no. 3 (1998): 17–32.

Corral Sánchez-Cabezudo, Francisco. "Presentación de la colección." In *Obras completas*, vol. 2, by Rafael Barrett; edited by Francisco Corral-Cabezudo, 19–20. Tantín, 2010.

Darwin, Charles. *The Descent of Man*. In *The Four Great Books of Charles Darwin*, edited by Edward O. Wilson. Norton, 2005.

Deleuze, Gilles. "The Actual and the Virtual." Translated by Eliot Ross Albert. In *Dialogues II: Gilles Deleuze and Claire Parnet*, 148–52. Columbia, 2007.

Deleuze, Gilles, and Felix Guattari. *A Thousand Plateaus: Capitalism and Schizophrenia*. Translated by Brian Massumi. University of Minnesota, 1987.

Delgado, Leandro. Prólogo a *Crónicas de la naturaleza*, by Rafael Barrett; edited by Leandro Delgado, vii–xxxi. Biblioteca Artigas, 2012.

Ellenberger, Henri F. *The Discovery of the Unconscious: The History and Evolution of Dynamic Psychiatry*. Basic Books, 1970.

Escobar, Ticio, and Osvaldo Salerno. "El grabado de Cabichuí como expresión popular." In *Cabichuí, periódico de la guerra de la triple alianza*, edited by Ticio Escobar and Osvaldo Salerno. Museo del Barro, 1984.

Ferretti, Federico. "Evolução e revolução: Os geógrafos anarquistas Elisée Reclus e Pëtr Kropotkin e sua relacao com a ciencia moderna, séculos XIX e XX. *Historia, Ciencias, Saúde*. SciELO Brazil, August 2017. https://www.scielo.br/j/hcsm/a/9GFD7p7dRBBVYZvGC5KZ6Qm.

Foucault, Michel. *The Birth of Biopolitics: Lectures at the College de France, 1978–79*, edited by Michel Senellart, translated by Graham Burchell. Palgrave Macmillan, 2008.

Foucault, Michel. *The Order of Things: An Archaeology of the Human Sciences*. Random House: 1970.

French, Jennifer L. "Yerba." In *Latin American Literature in Transition, 1870–1930*, edited by Fernando DeGiovanni and Javier Uriarte, 89–102. Cambridge, 2023.

French, Jennifer, and Gisela Heffes. "Introduction: Genealogies of Latin American Environmental Culture." In *The Latin American Ecocultural Reader*, edited by Jennifer French and Gisela Heffes, 3–14. Northwestern University Press, 2021.

French, Jennifer, and Gisela Heffes. "Developmentalism." In *The Latin American Ecocultural Reader*, edited by Jennifer French and Gisela Heffes, 211–14. Northwestern University Press, 2021.

Giorgi, Gabriel. Review of *Before the Law: Humans and Animals in a Biopolitical Frame*, by Cary Wolfe. *Emisférica* 10, no. 1 (Winter 2013). https://hemisphericinstitute.org/en/emisferica-101/10-1-book-reviews/giorgi.html.

Giorgi, Gabriel. *Formas communes: Animalidad, cultura, biopolítica*. Eterna Cadencia, 2014.

Glick, Thomas F., Rosaura Ruiz and Miguel Ángel Puig-Samper, eds. *El darwinismo en España e Iberoamérica*. Universidad Nacional Autónoma de México/Doce Calles, 1999.

Grosz, Elizabeth. *Becoming Undone: Darwinian Reflections on Life, Politics, and Art*. Duke University Press, 2011.

Grosz, Elizabeth, and Simone Stirner. "All Too Human: A Conversation with Elizabeth Grosz." *Qui Parle: Critical Humanities and Social Sciences* 25, no. 1-2 (Fall/Winter 2016): 17–33.

Hale, Charles. "Political Ideas and Ideologies in Latin America, 1870–1930." In *Ideas and Ideologies in Latin America*, edited by Leslie Bethell, 133–206. Cambridge, 1996.

Haraway, Donna J. *Staying with the Trouble: Making Kin in the Chthulucene*. Duke University Press, 2016.

Hoyos, Héctor. *Things with a History: Transcultural Materialism and the Literatures of Extraction in Contemporary Latin America*. Columbia University Press, 2019.

Jackson, Zakkiyah Iman. "Animal: New Directions in the Theorization of Race and Posthumanism." *Feminist Studies* 39, no. 3 (2013): 669–85.

Jackson, Zakkiyah Iman. *Becoming Human: Matter and Meaning in an Antiblack World*. New York University Press, 2020.

Kropotkin, Peter. *Mutual Aid*. Critical Editions, 2021.

Lawlor, L. "Henri Bergson." *Stanford Encyclopedia of Philosophy*, 2004. Online at https://plato.stanford.edu/entries/bergson/#CreaEvol.

Mignolo, Walter. "Commentary." In *Natural and Moral History of the Indies: José de Acosta*, edited by Jane Mangan, 451–518. Duke University Press, 2002.

Moore, Jason W. *Capitalism in the Web of Life: Ecology and the Accumulation of Capital.* Verso, 2015.

Muñoz, Vladimiro. *El pensamiento vivo de Rafael Barrett.* Rescate, 1977.

Pearson, Keith Ansell. *Germinal Life: The Difference and Repetition of Deleuze.* Routledge, 1999.

Pearson, Keith Ansell, and John Ó Maoilearca. Introduction to *Key Writings*, by Henri Bergson; edited by Keith Ansell Pearson and John Ó Maoilearca, 1–56. Bloomsbury Academic, 2014.

Plá, Josefina. "El grabado, instrumento de la defensa." In *Cabichuí, periódico de la guerra de la triple alianza*, edited by Ticio Escobar and Osvaldo Salerno. Centro de Artes Visuales - Museo del Barro, 1984.

Pratt, Mary Louise. *Imperial Eyes: Travel Writing and Transculturation*, 2nd ed. Routledge, 2008.

Pratt, Mary Louise. *Planetary Longings.* Duke University Press, 2022.

Quijano, Aníbal. "Coloniality of Power, Eurocentrism, and Latin America." In *Coloniality at Large: Latin America and the Postcolonial Debate*, edited by Mabel Moraña, Enrique Dussel, and Carlos A. Jáuregui, 181–224. Duke University Press, 2008.

Reclus, Elisée. *Anarchy, Geography, Modernity: Selected Writings.* Edited and translated by John Clark and Camille Martin. PM Press, 2013.

Roa Bastos, Augusto. "Rafael Barrett: Descubridor de la realidad social del Paraguay." Prologue to *El dolor paraguayo*, by Rafael Barrett. Biblioteca Ayacucho, 1978.

Schmitt, Cannon. *Darwin and the Memory of the Human: Evolution, Savages, and South America.* Cambridge University Press, 2009.

Vara, Ana María. *Sangre que se nos va: Naturaleza, literatura y protesta social en América Latina.* Consejo Superior de Investigaciones Científicas, 2013.

Viñas, David. *Anarquistas en América Latina.* Paradiso, 2004.

Villalba Peña, Rodrigo N. "Rosicrán y su poema Ñande ỹpỹ cuéra (1929)." *Boca del sapo* 33, no. 23 (February 2022): 38–51.

Viveiros de Castro, Eduardo. "Cosmological Deixis and Amerindian Perspectivism." *Journal of the Royal Anthropological Institute* 4, no. 3 (Sept. 1998): 472.

CHAPTER 11

Graffiti as Earthly Inscriptions

Human Acts and Geological Forces in Euclides da Cunha's Os sertões (1902)

EMMANUEL A. VELAYOS LARRABURE

Translating Scales

In the 1901 article "Fazedores de desertos" (Desert makers), the Brazilian savant and journalist Euclides da Cunha (1866–1909) discussed how farming practices had altered the climate of southern Brazil, producing droughts and deserts in an otherwise tropical region.[1] Like several other nineteenth-century Brazilian intellectuals, da Cunha denounced the human-led destruction of nature.[2] A civil and war engineer by training, the polymath da Cunha did not lack the tools such a stance required: his writings explored a wide array of scientific and aesthetic discourses, including geography, geology, literary travelogues, and phrenology. During the 1890s, moreover, he developed first-hand knowledge of the topography of southern Brazil through his work as a public servant: first as an officer in the Public Works Department of the State of São Paulo and later as a member of the Instituto Histórico e Geográfico de São Paulo. In "Fazedores de desertos" he used that knowledge to link Indigenous and Western farming practices with a Brazilian tradition of swidden cultivation and desertification.[3] Those practices were associated with the northern backlands, yet the text claims they also affected southern states, like São Paulo. But more than being a regional or national problem, slash-and-burn agriculture and its effects on the climate would bespeak the growing human-led impact on the environment. They showed, for da Cunha, that humans were "um agente geológico nefasto, e um elemento de antagonismo terrivelmente bárbaro da própria natureza que nos rodeia" (a nefarious geological agent, an element of terribly barbarous antagonism to the very nature that surrounds us).[4]

The article does not directly discuss the implications of such a geological agency but suggests what it entails: "O homem, a quem o romântico historiador negou um lugar no meio de tantas grandezas [naturais], não as corrige, nem as domina nobremente, nem as encadeia num esforço consciente e sério. Extingue-as" (Man, to whom the romantic historian denied a place amid natural wonders, does not correct them, nor dominate them nobly, nor chain them in a conscious and serious effort. Man extinguishes them).[5] Romantic writers conceptualized nature as a realm governed by stable patterns over which humans had no control; for da Cunha, humans had not gained control over such patterns but were annihilating them. In his view, humans had not become ecological actors within larger natural processes; they had grown into an extreme force of environmental destabilization. They were shattering entire ecosystems, as happens when the planet's deep history activates in the course of major geological transformations, producing mass extinctions. It is to rhetorically highlight such "nefarious" effects of human action that da Cunha placed it beyond the biological/ecological realms and within the scale of geological processes.[6]

Da Cunha's rhetorical strategy anticipates the work of Eugene F. Stoermer and Paul Crutzen, the ecologist and atmospheric chemist (respectively) responsible for the coinage and dissemination of an important contemporary neologism: the Anthropocene. As Dipesh Chakrabarty and others have subsequently argued, postulating humans as a collective geological "agent" implies a translation from the physical field of geological forces into the anthropocentric and social realm of agency, a translation with major moral implications. Geological forces operate within the vast temporal scales of planetary history, which challenge our imagination: they are too large to be apprehended in terms of human history or through a political or policy-oriented lens. The translation of force into agency circumvents this impasse: it allows us to develop a moral/political consciousness of our responsibility for climate change.[7] In this vein, while da Cunha stated that humans did not control their impact on the environment, his use of the term "agent" underlines how he attributed the responsibility for Brazilian climate change to a concrete human activity: slash-and-burn farming. And in fact the denunciation of this practice continued an ecological agenda within da Cunha's career: during this time in the Public Works Department and the Instituto Historico e Geografico, he fought against swidden agriculture and advocated for replacing it with sustainable irrigation and better land management.[8]

Da Cunha's most important book, *Os sertões* (1902; Backlands), also portrays humans as a collective geological agent to emphasize the magnitude of their environmental impact. The section "Como se faz un deserto" (How to make a desert) discusses several factors producing deserts in northeastern Brazil, among them "um agente geológico notável–o homem. Este, de fato, não raro reage brutalmente sobre a terra e . . . assumiu, em todo o decorrer da história, o papel de um terrível fazedor de desertos" (one notable geological agent–man. In fact, he most usually has had a brutal effect upon the land and . . . has assumed along the whole part of history the role of terrible maker of deserts).[9] However, the 1901 piece and the book have different approaches to the geological agency of humans.[10] The article does not fully elaborate on its geological implications; it rather privileges the anthropogenic realm of responsibility and agency. The book explores the relationship between geological and human realms in a more complex fashion: instead of just focusing on the one-sided translation of geological forces into human agency, it also engages in the conversion of human dynamics into geological ones.

The present chapter studies how this two-sided translation in *Os sertões* combines the moral structures of the human realm with the scales of geology and the earth's deep time. I argue that the primary means of linking geology to humans in the book is the overlap between two forms of rhetoric. On the one hand, the book is permeated with a geological rhetoric rendering every natural and social formation as a product of the superposition of strata from different ages of the earth. On the other hand, there is a textual imagination depicting landscapes, earth strata, and humans as the palimpsestic overlay of heterogeneous manuscripts brought together in a compulsive, enigmatic, and never-ending rewriting. By translating earthly dynamics into the realm of the written word, the palimpsestic rhetoric brings geology within the grasp of human and affective structures, while the depiction of landscapes and humans through the constant reference to an excessive, inscrutable textuality maintains the irreducible character of geological forces. Conversely, the geological rhetoric inserts human dynamics and politics within the layers of immemorial and anonymous geophysical processes. As we will see, the overlap between these two forms of rhetoric is at play in an exemplary passage where human action is portrayed in both geological and textual terms: a communal graffiti composed of several layers—or strata—of palimpsestic wall inscriptions.

Geological Strata and Palimpsests

The overlay between human dynamics and geological forces is exhibited in the book's composition. Initially, *Os sertões* was to assemble the chronicles da Cunha penned as a correspondent covering the War of Canudos (1897), where the army of the Republic of Brazil decimated a subaltern uprising headed by Antônio Conselheiro (1830–1897). His early chronicles celebrated the heroism of the military; however, as the writer began to question the morality of the conflict, he came to regard it as a state crime and rewrote his texts to explore deeper geological causes to which he attributed the war.[11] The outcome combined the geological treatise with geographical surveys, aesthetic travelogues, and phrenological and racial reflections.[12] The first part of the book, "A terra" (The land), is a geological and aesthetic survey of the deep earth's dynamics that, according to da Cunha, controlled the climate in northeastern Brazil. The second part, "O homem" (Man), explores how such geological factors had determined the racial and moral composition of the region's inhabitants. The third part, "A luta" (The battle), chronicles the war with the same geological rhetoric at play in the previous parts.

Overall, the book presents human actions in geological terms, that is, as anonymous and immemorial processes exceeding the scope of intentionality and free will. This dismissal of human agency could be due to the pervasive environmental determinism in positivist sciences. However, the book explores geological discourse with a high degree of eclecticism and proposes fluid links between the spheres of earth's deep forces and human agency beyond any straightforward determinism. This eclecticism is related to da Cunha's take on the politics of positivist science in Brazil: as he grew critical of the state massacre in Canudos, he became aware of the positivist intelligentsia's complicity in it. In the book, he claimed that by using European scientific and civilizational ideals to justify the state violence, Brazilian intellectuals had played the role of "mercenários inconscientes" (unconscious mercenaries).[13] Given this self-criticism, it is not surprising that *Os sertões* is peppered with a skeptical stance on the explanatory potential of modern science. Da Cunha even stated that his use of geological and racial terms was meant to explore "todas as variáveis de uma fórmula intricada, traduzindo sério problema; mas não desvendamos todas as incógnitas" (all the variables of a complex formula, translating it into a complex problem, yet without solving its unknowns).[14] Rather than merely throw light on his subject matter, da Cunha's scientific discourse highlights its intricate nature.

Along these lines, "A terra" depicts the Brazilian backlands as a succession of chaotic geological formations that came about from the manifold juxtaposition of strata from different times in the earth's deep history. As da Cunha puts it, in the northeastern countryside, "formações geognósticas díspares, de idades mal determinadas, . . . se substituem, ou se entrelaçam, em estratificações discordantes, formando o predomínio exclusivo de umas, ou a combinação de todas" (quite dissimilar geological formations of indeterminate age supplant each other or intermingle in a discordant stratification, which makes for the exclusive predominance of some or a combination of all).[15] These formations would result from constant friction between different earth strata animated by the inner onslaught of geological forces. He considered that such geological dynamics produced "o martírio da terra, brutalmente golpeada pelos elementos variáveis" (the martyrdom of the land, brutally lashed by all the different elements).[16] This violent struggle had traversed millennia and could only be perceived from a viewpoint encompassing the earth's different ages at the same time: "[É uma] luta surda, cujos efeitos fogem ao próprio raio dos ciclos históricos, mas emocionante, para quem consegue lobrigá-la ao, através de séculos sem conto" (It is a mute battle, the effects of which are far removed from the normal course of historical cycles, although in a rather moving way for someone who manages to catch sight of it over the countless centuries).[17] Such an all-encompassing viewpoint exceeded the chronologies of anthropocentric history and was unattainable to the human eye.

However, for da Cunha, the human eye could still grasp the aesthetic effects of the inner geological struggle, for they produced "[uma] plasticidade admirável aos mais caprichosos modelos" (a wonderful plasticity as they take on the most capricious forms).[18] It was by depicting the "plastic" overlap of various strata that the author staged a trans-historical, earth-centered perspective from which to appreciate the results of the mute geological battle. As his accounts were not organized in terms of background and foreground, they did not produce a coherent perspectival effect through which the observer could control the landscape's representation.[19] Indeed, rather than being organized as perspectival paintings, his geological landscapes were structured as a vertical juxtaposition of different layers on the same point: "em plano vertical, sucedendo-se a partir da base . . . em alongado roteiro pela superfície" (One after the other, on a vertical plane from their base . . . in a lengthy route along their surface).[20]

The observer's impressions did figure in these depictions, but only to signal how overwhelming that loaded overlapping was for the human gaze. After traversing the peaks surrounding Canudos, the author asserts that "o observador que seguindo este itinerário deixa as paragens . . . ao atingir aquele ponto estaca surpreendido" (an observer who has been following this route is now leaving the landscapes . . . when he comes to that point he will stop short in surprise).[21] Likewise, reproducing the observer's wonder when entering Canudos, he states that "[é] uma paragem impressionadora. As condições estruturais da terra lá se vincularam à violência máxima dos agentes exteriores para o desenho de relevos estupendos" (It is an impressive landscape. The structural makeup of the land has been coupled with a great upheaval of external agents in the design of stupendous reliefs).[22] Luciano Tosta has noted that *Os sertões*'s depiction of nature goes hand in hand with a detailed account of the amusement and bewilderment it produces in the observer.[23] The dismantling of the perspectival landscape produces in the viewer an affective experience of bewilderment that is intellectually exacerbated by what the overlap of geological strata unveils: earth's temporalities exceeding our capacity of conceptualization.[24] Along these lines, Carlos Fonseca has described da Cunha's geological accounts as "radical landscapes," refusing to be incorporated into the political mechanisms of representation that characterize modern science.[25] These "radical landscapes" can bring the temporal scales of geological time within the grasp of human structures, but only by undoing the human-centered perspective to control the backlands' representation. This disruptive geological rhetoric also informs the ethnic and racial reflections in other parts of the book—including, for example, where da Cunha describes the ethnic features of Antônio Conselheiro and his followers ("conselheristas")

> como os estratos geológicos não raro se perturbam, invertidos, sotopondo-se uma formação moderna a uma formação antiga. . . . A estratificação moral dos povos por sua vez também se baralha, e se inverte, e ondula riçada de sinclinais abruptas, estalando em *flaults*, por onde rompem velhos estádios há muito percorridos.[26]

like geological strata, which when disturbed or inverted reveal a modern formation below an ancient one. . . . The moral stratification of people can also show an inversion and melding of layers, with sinuous furrows and abrupt synclinal eruptions that break into faults in the form of ancient strata through which the race has long passed.[27]

This passage vividly illustrates how the geological imagination of multilayered strata exceeding a representational viewpoint permeates the account of the backlands' inhabitants. Just as the overlap of remote and modern geological layers puts strata from different earth-ages in direct contact, the racial and moral constitution of the conselheristas is laden with an assortment of heterogeneous elements. Adriana Campos Johnson has argued that the backlands, not their inhabitants, were the book's central protagonist, which turned the depiction of these subjects into a kind of geographical representation.[28] I would add that rather than determined by the cartographic planes of modern geography, da Cunha's take on the conselheristas is informed by the geologist's unearthing drive. In this sense, he approaches the behavior of that leader "da mesma forma que o geólogo, interpretando a inclinação e a orientação dos estratos truncados de antigas formações, esboça o perfil de uma montanha extinta" (just as the geologist reconstructs the inclination and orientation of very old formations from truncated strata and builds the model of ancient mountains).[29]

This comparison could lead to scientific determinism in the depiction of conselheristas, but in reality it has a rather eclectic meaning. Instead of leading to a fixed view of the planet's past, the deep and clashing temporalities unearthed by nineteenth-century geological discourse "showed that the Earth's deep history—and therefore its future—could not be reduced to any . . . simple or predictable form."[30] In this respect, just as da Cunha unleashes geological discourse's disruptive potential against perspectival landscapes, his geological take on the leader of Canudos does not follow a unifocal perspective to render the relationship between geological-environmental forces and human agency. The section titled "Antônio Conselheiro, documento vivo de atavismo" (Antônio Conselheiro, a Living Document of Atavism), represents the leader as a palimpsest conveying conflating temporalities and, at the same time, illustrates how da Cunha imagines a complex interaction between the leader and his environment:

Ele foi, simultaneamente, o elemento ativo e passivo da agitação de que surgiu. O temperamento mais impressionável apenas fê-lo absorver as crenças ambientes, a princípio numa quase passividade pela própria receptividade mórbida do espírito torturado de reveses, e elas refluíram, depois, mais fortemente, sobre o próprio meio de onde haviam partido, partindo da sua consciência delirante.[31]

He was, at the same time, a passive and an active agent of the rebellion that surged up around him. His highly impressionable temperament led him to absorb the state of mind of his environment. At first, his tormented mind was little more than a vessel. With time his delirious consciousness unleashed these forces on the surroundings that had produced him.[32]

The geological take on Conselheiro's life recognizes complex interactions between him and his environment that exceeded any fixed viewpoint. By embodying environmental processes in the leader's persona, the passage helps grasp geological forces on a human scale.[33] Yet, in signaling a degree of active agency for the Conselheiro, it also counterbalances the possibility of environmental determinism. Just like the overlap of geological strata refuses any fixed or stable perspective in *Os sertões*, the book's take on the dynamics between Conselheiro and the backlands does not reduce either to the other.

The depiction of the Brazilian backlands and their inhabitants as multilayered overlaps conveyed an experimental attempt at translating and crossing geological and human scales. We find a similar imagination of stratified inscriptions in da Cunha's take on the conselheristas' heterogeneous religious beliefs, of which da Cunha stated that "como um palimpsesto, a consciência imperfeita dos matutos revela nas quadras agitadas, rompendo dentre os ideais belíssimos do catolicismo incompreendido" (as in a palimpsest, the imperfect consciousness of the backlander reveals itself during turbulent seasons, shattering the beautiful ideas of a misplaced Catholicism).[34] Just as the overlapping of geological strata conveys several stages of the earth's past, the metaphor of an overloaded palimpsest contains beliefs from various times in human history. The palimpsest image translated the geological rhetoric into the human domain of the written word so that the multiplicity of earth's temporalities could be grasped from a human perspective. Yet this perspective does not reduce geological processes to human agency, for the palimpsestic dynamics would still be animated by tumultuous environmental forces.

Da Cunha developed his palimpsestic imagination further in one of the means through which he depicted the enigmatic persona of Conselheiro: the portrayal of his handwriting. In a passage on the vestiges that the soldiers picked from the conselheristas, the author focuses on their leader's manuscripts. Instead of discussing the content of those texts, da Cunha underscores his perplexity at their disconcerting visuality and style: "Pobres papéis, em que a ortografia bárbara corria parelhas com os mais ingênuos absurdos e a escrita irregular e feia parecia fotografar o pensamento torturado, eles resumiam a psicologia da luta" (The limp sheets, with their irregular handwriting and barbarous spelling, seemed to be a photographic image of twisted thoughts. . . . They summarized the psychology behind the conflict).[35] If Conselheiro's character is an intricate "text" conflating several arcane layers of deep geological history, his manuscripts provide a photographic depiction of that intricate overlapping. There is a vibrant dimension in how the author deploys these visual tropes to depict Conselheiro as a loaded palimpsest and portray his handwriting as a manifestation of that overwritten text. Far from any bookish inertia, that metaphoric textuality would be animated by an active physical and bodily drift like the inner dynamism in the intense juxtaposition of geological layers in "A terra." Paralleling that incessant geological struggle, da Cunha pictures Conselheiro's handwriting as a convulsive rendition of an internal battle in the leader's soul.

Graffiti: Outrageous Palimpsests

With the rise of graphological studies in the second half of the nineteenth century, handwriting was increasingly seen as a mirror of an individual's intellectual and moral qualities.[36] In his take on Conselheiro's manuscripts, da Cunha combined this heuristic conceptualization of handwriting with the indexical power of photography to represent reality. Reflecting on the rendition of such manuscripts, Rachel Price noted that "these purportedly indexical—photographic—writings are taken to be both unmediated expressions and allegories of the conflict: the outward expression of interiority and the reproduction of the outward aspect of the conflict."[37] Yet, in the book, this sense of immediacy is counterbalanced by inner textual dynamics turning every account of Canudos into one more layer of writing in an overloaded array of textual strata.[38] Os sertões is constructed as a textual palimpsest redrafting and reinterpreting previous accounts of the conflict and producing new versions out of them. Indeed, da Cunha himself emphasized the significant difference between this rewriting and his first journalistic pieces, for the final text exhibited, in his words, "outra feição, tomando apenas variante de assunto geral o tema, a princípio dominante, que o sugeriu" (a different form. The theme that had been the main one at the start and was to give the work its inspiration is now but one of the many dealt with here).[39]

Moreover, in the book, da Cunha approaches the multifarious rewriting of different textual and visual renditions of Canudos through the metaphor of "palimpsestos ultrajantes."[40] The phrase appears in "A Luta" to depict the unruly intermingling of several layers of dissident soldiers' writings on a group of walls. The "outrageous palimpsests" in the third part echo the geological rhetoric of the superposition of layers in the previous parts of the book and reflect the overlap between geological and human scales. A keyword of manuscript media, "palimpsest" encapsulates, in a synchronic image, the several diachronic layers of writing and redrafting that compose a handwritten document. The term evokes a stratification of textual layers similar to the overlap of geological strata, yet the corporeal connotations of manuscript media frame such stratification within a human scale. As the term alludes to the hand's traces in the recycling of manuscripts, it projects an embodied and human connotation onto the overlap and rewriting of different sorts of textual modalities in da Cunha's book. In addition, "palimpsest" has been used as a metaphor to trace the multiple and changing ideological positions in the subject of enunciation of a text, while the qualifier "outrageous" adds a chaotic and unsettling meaning to the overlap of different textual and ideological layers in *Os sertões*.[41] Considering all the above, the phrase "outrageous palimpsests" suggests that da Cunha rewrote his initial war chronicles from a different ideological perspective, one in which the war is not a glorious deed so much as a state massacre. This change of attitude is exemplarily evidenced when contrasting the different textual accounts that da Cunha penned regarding the soldiers' graffiti.

His first take on those wall inscriptions was published in a war chronicle of September 1, 1897. There, da Cunha recounted having observed in the town of Queimadas, "nas paredes brancas, sobre a brancura da cal, a traços de carvão, numa caligrafia hieroglífica, ostenta-se a verve áspera e característica dos soldados" (on the white walls, over the whitewash of lime, and with traces of charcoal, the rugged verve characteristic of soldiers in hieroglyphic calligraphy).[42] Combining manuscript and war metaphors to depict that graffiti as collective writing by armed men, he adds that "todos os batalhões colaboraram na mesma página; uma página demoníaca: períodos curtos, incisivos, assombrosos, arrepiadores, espetados em pontos de admiração maiores do que lanças" (all the battalions collaborated on the same page, a demonic page: short, insidious, shocking, violent, lapidary passages surrounded by exclamation marks bigger than spears).[43] Despite such textual metaphors, the article does not elaborate further on the graffiti's content. Moreover, while the wall inscriptions resulted from several layers of writing, da Cunha does not present them through a geophysical imagination of strata. Instead, the chronicler approaches the graffiti here as a collective gesture configuring an affective landscape: "Impressiona a passagem pelo lugar onde se agitaram tantas paixões e se acalentaram tantas esperanças malogradas" (It is impressive to pass through a site containing so much passion and lost hope).[44]

Da Cunha was far from celebrating the "demonic" tone of those wall inscriptions, yet he recorded his fascination in looking at them. He detected an abrupt and violent dimension in this graffiti, but he linked it to warfare's typical violence: for him, those soldiers inscribed exclamation marks on walls like they launched spears at the enemies on the battlefield. Since the chronicle narrates how the army advanced toward the conflict zone, the vivid depiction of these marks registers the gestural means by which representatives from all battalions took possession of the backlands' space. This graffiti would be a sort of communal inscription through which the army claimed the land of Canudos for the authority that they represented and that sent them to the conflict: the republic. By documenting this graffiti in his chronicle, da Cunha provided a public audience for the political and military gestures through which the republic won the conflict of Canudos. Hence, as the Brazilian writer was a supporter of the republic at that time, he showed some indulgence toward this graffiti in his chronicle.

However, in rewriting this episode in *Os sertões*, he was much more severe in accounting for these "demonic pages." He ironically described the graffiti with penmanship and lettered metaphors, asserting how he witnessed "nas suas paredes, cabriolando doidamente, a caligrafia manca e a literatura bronca do soldado" (on the walls, stepping on each other, the soldiers' lame handwriting and scolding literature).[45] He also stressed these wall inscriptions' collective and violent character by recalling that "todos os batalhões haviam colaborado nas mesmas páginas, escarificando-as à ponta de sabre ou tisnando-as a carvão, no gravarem as impressões do momento. Eram páginas demoníacas" (all battalions had collaborated on the same pages, scarifying them with the tip of their sabers or charcoal to record their impressions. They were demonic pages).[46] He elaborates further on the "demonic" character of such graffiti by mentioning some formal elements of that collective handwritten expression: "períodos curtos, incisivos, arrepiadores; blasfêmias fulminantes; imprecações, e brados, e vivas calorosos, rajavam-nas em todo o sentido, profanando-as, mascarando-as, em caracteres negros espetados em pontos de admiração, compridos como lanças" (short, incisive, chilling passages fulminating blasphemies, curses, cries, and warm cheers: [words] blew in every direction, desecrating and overwriting one another, in black characters spiked with exclamation marks long as spears).[47]

In the chronicle, da Cunha implicitly condones these indecent inscriptions as a vigorous gesture whose excessive character is typical of the atmosphere of war. But he does not spare space to describe those graffiti as blasphemies, insults, and profanities in the book. Since he was no longer supporting the "glorious" actions of the republican army, he revises his initial take on these "demonic pages" and renders them as a surplus of violence and indecency. Thus while there are textual similarities between the chronicle's fragment and its rendition in "A Luta," this passage shows how the author rewrites his war chronicles in the book from a different ideological perspective. Note that this new take on those wall inscriptions parallels the account of Conselheiro's handwriting in the book: both forms of writing are portrayed as embodied texts whose stylistic and visual features materialize an excessive dimension. By the same token, in the book, da Cunha adds another rendition of those writings on the wall, one in which he regards them as "photographic" versions of the conflict—just like he did with Conselheiro's handwriting. This additional account of the graffiti takes place some pages after the rewriting of the chronicle's fragment in which he recounts having witnessed a dissident performance by a group of wounded soldiers:

Em cada parede branca . . . se abria uma página de protestos infernais. Cada ferido, ao passar, nelas deixava, a riscos de carvão, um reflexo das agruras que o alanceavam, liberrimamente, acobertando-se no anonimato comum. A mão de ferro do exército ali se espalmara traçando em caracteres enormes o entrecho do drama; fotografando, exata, naquelas grandes placas, o facies tremendo da luta em inscrições lapidares, numa grafia bronca, onde se colhia em flagrante o sentir dos que o haviam gravado.

Sem a preocupação da forma, sem fantasias enganadoras, aqueles cronistas rudes deixavam por ali, indelével, o esboço real do maior escândalo da nossa história—mas brutalmente, ferozmente, em pasquinadas incríveis—libelos brutos, em que se casavam pornografias revoltantes e desesperanças fundas, sem uma frase varonil e digna. A onda escura de rancores que rolava na estrada chofrava aqueles muros, entrava pelas casas dentro, afogava as paredes até ao teto . . . A comitiva, penetrando-as, repousava envolta num coro silencioso de impropérios e pragas . . . E a empresa perdia repentinamente a feição heróica, sem brilho, sem altitude. Os narradores futuros tentariam em vão velá-la em descrições gloriosas. Teriam em cada página, indestrutíveis, aqueles palimpsestos ultrajantes.[48]

On each white wall, there was a page of infernal protests. Hidden in their common anonymity, wounded soldiers wrote them with pieces of coal to express the hardships that tormented their souls. The iron hand of the army opened there, tracing the drama of the conflict in giant characters: photographing, on those large surfaces, the tremendous *facies* of the struggle in lapidary inscriptions, in a bold script recording its writers' feelings.

Without any formal concern, without deceptive fantasies, those rude chroniclers left, indelibly, the real outline of the greatest scandal in our history. They did so brutally, ferociously, in incredible screeching. Their writings were gross libels, in which revolting pornography and deep hopelessness mingled without any dignified sentence. The dark wave of grudges that rolled on the road buffeted those walls, entered the houses, and drowned the walls up to the ceiling . . . As the entourage entered the houses, they rested in a silent chorus of profanities and curses . . . And the military endeavor suddenly lost its heroic appearance. Future narrators would try in vain to veil it in glorious descriptions. They would have on each page, indestructible, those outrageous palimpsests.[49]

This new layer of writing and meaning that da Cunha adds to his array of accounts on war graffiti marks a more radical departure from his indulgent take on the soldiers' "demoniac pages" in the 1897 chronicle. There is no longer a group of armed men from all the battalions of the army deploying their writing to mark the state's authority over a conquered space; instead, we find a group of wounded soldiers inscribing several layers of outrageous and pornographic messages that scorn their commanders. Even more outrageous than the messages themselves is their visual disposition: the overlapping of texts in which profane cheers and boos mix. Before the narration of this episode, it is stated that these wounded men heralded the official convoy of General Arthur Oscar, the military chief of the army in Canudos. In passing near those wall inscriptions, the General and his convoy would have felt suddenly repelled by a chorus of insults and profanity. In sum, what the journalistic pieces depicted as a gesture of collective possession of the space by a unified republican army is resignified here as a dissident performance in which a subaltern group of soldiers displays their angry complaints against their leaders. This new version would depict one of the most important institutions of the republic as a divided group whose leaders were exposed to internal criticism.

It is telling that the author regards the wounded soldiers as the truest "chroniclers" of the conflict. In my view, in that meta-textual reflection, da Cunha imagines the task of chronicling Canudos as an incessant rewriting of previous texts and discourses, a relentless reworking that cannot be subsumed by any fixed political or scientific modality. Rather than reproduce a coherent ideological or scientific discourse, that chronicling deploys the written word as a passionate means to develop graphic and embodied images conveying an experimental geological depiction of Canudos and the country's political and cultural reality. If these textual dynamics do not lead to a fixed perspective, such "outrageous palimpsests" are animated by an incessant and disruptive dynamism comparable to the inner geological struggle that, according to da Cunha, affects the backlands and their peoples: the mute battle that transcends the War of Canudos.

The author uses the term "facies"—the external aspect of sedimented rocks—to describe those overwritten walls, which projects a geological connotation onto them. Moreover, these tumultuous wall inscriptions are portrayed as being animated by a hectic and impromptu movement of words resembling the capricious movement of geological strata inside the land. They would also mirror da Cunha's account of Conselheiro's handwritings: like that leader's texts, this graffiti would express visually and stylistically the telluric forces producing the conflict. Hence the "outrageous palimpsests"

show the rhetorical connivance between geological and palimpsestic metaphors that brought together earth's deep time and human actions.

In projecting overlaps of geological strata upon human dynamics and
inscriptions (graffiti), da Cunha posited human acts as anonymous and
immemorial geophysical forces exceeding the human scope of intentionality and free will. Such disdain for human freedom could gesture toward
the geographical determinism of positivist science. However, the book
renders environmental and human dynamics as tangled forces exceeding
the anthropocentric domain of intentionality and epistemic operations
of appropriation and control. Indeed, in *Os sertões*, human and planetary
scales intertwine: the geological rhetoric explores the manifold and disruptive temporalities of earth's deep time, while palimpsestic imagination
translates geological and environmental processes into the human realm
of the written word.

This interweaving of scales resonates with current epistemological and
political debates on the Anthropocene. Nigel Clark notes that besides the
conversion of geological terms into human/political ones, the Anthropocene requires the reverse process: to resituate political concerns within a
geo-temporal frame "that radically exceed[s] any conceivable human presence."[50] This reconversion should not replace anthropocentric approaches
with entirely geocentric or deep ecology ones, perspectives spurring much
criticism for their lack of concern for human inequalities.[51] Instead, we
should explore links between geological and human scales without entirely
reducing either to the other.[52] In this fashion, the two-fold translation of
human dynamics and earthly processes in *Os sertões* produces a chiasmus that,
in Chakraborty's terms, "brings within the grasp of affective structures of
human-historical time the vast scales of geobiology that these structures do
not usually engage."[53] In helping us imagine geophysical dynamics within
our affective and moral structures, this chiasmus of scales fosters a geolog-
ically informed politics in the Anthropocene. The possibility of that politics has inspired my reading of the book.

NOTES

Support for this project was provided by a PSC-CUNY Award, jointly funded by The Professional Staff Congress and The City University of New Yor1.

1. Cunha, "Fazedores de desertos"; my translation.

2. See Nielson, "Nefarious Agents of Ecological Collapse," 1–5.
3. With regard to such farming, da Cunha stated "tais selvatiquezas atravessaram toda a nossa história" (such savagery has gone through our entire history). Da Cunha, "Fazedores de desertos"; my translation.
4. da Cunha, "Fazedores de desertos," 1; my translation.
5. da Cunha, "Fazedores de desertos," 1; my translation.
6. On the contrast between the geological and biological scales of human agency, see Chakrabarty, "The Climate of History," 206.
7. Such an overlap of geological and moral realms informs many approaches to the Anthropocene, which tend to focus on the "displacement-translation of the [geological] category of 'force' . . . into the human-existential category of power and its sociological-institutional correlates." Chakrabarty, "Anthropocene Time," 9.
8. On da Cunha's ecological agenda, see Nielson, "Nefarious Agents of Ecological Collapse," 7.
9. da Cunha, *Obra completa*, 39; *Backlands*, 48. The English versions of the text are quoted from the Penguin edition.
10. Nielson claims the article's take on the human moves in the opposite direction of *Os sertões*, which presented humans as the result of environmental factors and not the other way around. (Nielson, "Nefarious Agents of Ecological Collapse," 11). On the depiction of the human in other da Cunha texts, see Nielson, "Ecological Thought in Euclides da Cunha's Correspondence and Writings about the Amazon," 17–34.
11. On the celebratory coverage of the war, see Galvão, *No calor da hora*. On how a large sector of the population did not share such a perspective, see Levine, *Vale of Tears*, 27. On how da Cunha developed a critical perspective on the morality of the war, see Galvão, *Saco de gatos*, 56.
12. Bernucci suggests that the book mixed two discursive modalities: one from sciences, the other from imaginary and literary realities. (Bernucci, *A imitação dos sentidos*, 117).
13. da Cunha, *Obra completa*, 99; my translation.
14. da Cunha, *Obra completa*, 147; my translation.
15. da Cunha, *Obra completa*, 102; *Backlands*, 12.
16. da Cunha, *Obra completa*, 111, 18.
17. da Cunha, *Obra completa*, 137, 47.
18. da Cunha, *Obra completa*, 100, 10.
19. With regard to the representation of landscapes in imperial travelogues, Mary Louise Pratt observed that "the sight is seen as a painting and the description is ordered in terms of background, foreground, . . . and so forth." Pratt, *Imperial Eyes*, 201.
20. da Cunha, *Obra completa*, 100; *Backlands*, 10.
21. da Cunha, *Obra completa*, 106, 12.
22. da Cunha, *Obra completa*, 111, 17.
23. Tosta, "O 'sublime' sertão em *O sertanejo*, *Os sertões* e *Grande sertão*," 7–26. Javier Uriarte has also noted that, in the book, "the sertão imposes its rhythms on the traveler." Uriarte, *The Desertmakers*, 216.
24. For the ways in which modern subjectivities can experience the end of perspectival landscapes as both a schizoid and emancipatory experience, see Andermann, *Tierras en trance*, 21–30.
25. Fonseca, *The Literature of Catastrophe*, 3. For Fonseca, da Cunha's "radical landscapes" challenge what Jens Andermann has called "the optic of the state," the scientific mech-

anisms of visualization of modern state-building. See Andermann, *The Optic of the State*.

26. da Cunha, *Obra completa*, 346–47.

27. da Cunha, *Backlands*, 281.

28. Campos Johnson, *Sentencing Canudos*, 135–36. Melo e Souza proposed the term "geopo-etics"–a poetics that emerges from the earth–to study da Cunha's geographical rhetoric. Souza, *A geopoética da Euclides da Cunha*, 7.

29. da Cunha, *Obra completa*, 203; *Backlands*, 124.

30. Rudwick, *Earth's Deep History*, 8.

31. da Cunha, *Obra completa*, 205.

32. da Cunha, *Backlands*, 125.

33. In this vein, in the aforementioned passage, the intriguing phrase "crenças ambientes" renders environmental forces as configuring a state of mind that is personified in the leader; this wording helps grasp geological forces on a physiological scale and, thence-forth, within the realm of the human.

34. da Cunha, *Obra completa*, 195; my translation.

35. da Cunha, *Obra completa*, 245; *Backlands*, 169.

36. Kittler, *Discourse Networks*, 262.

37. Price, *The Object of the Atlantic*, 104.

38. Even the use of pictures in the book was marked by textual dynamics of rewriting and reinterpretation that unearthed hidden meanings of the state massacre within the offi-cial photographs commissioned by the army. See Andermann, *The Optic of the State*, 199.

39. da Cunha, *Obra completa*, 99; *Backlands*, 1.

40. da Cunha, *Obra completa*, 457–58.

41. See Adorno, "Textos imborrables," 55–68.

42. da Cunha, *Canudos e outros temas*, 55; my translation.

43. da Cunha, *Canudos e outros temas*, 55; my translation.

44. da Cunha, *Canudos e outros temas*, 54; my translation.

45. da Cunha, *Obra completa*, 452; my translation.

46. da Cunha, *Obra completa*, 452; my translation.

47. da Cunha, *Obra completa*, 452; my translation.

48. da Cunha, *Obra completa*, 457–58.

49. My translation.

50. Clark, "Geo-politics and the Disaster of the Anthropocene," 27–28. On how to resituate the political within a geological time, see Clark, "Politics of Strata," 1–21. On the need to expand the field of inquiries of the Anthropocene beyond human-centered perspectives, see Zalasiewicz, "The Extraordinary Strata of the Anthropocene," 115–32.

51. Guha criticized the deep ecology movement for its "lack of concern with inequalities within human society." Guha, "Radical American Environmentalism and Wilderness Preservation," 72–74. However, an anthropocentric perspective is not enough to address global inequalities, for anthropocentrism is tied to the imperial ideologies producing such inequalities. Huggan and Tiffin, "Introduction," 5.

52. Along these lines, DeLoughrey and Handley propose a postcolonial ecology as "a com-plex epistemology that recuperates the alterity of both history and nature, without reduc-ing either to the other." DeLoughrey and Handley, "Introduction," 4.

53. Chakrabarty, "Anthropocene Time," 30.

BIBLIOGRAPHY

Adorno, Rolena. "Textos imborrables: Posiciones simultáneas y sucesivas del sujeto colonial." *Revista de Crítica Literaria Latinoamericana* 21, no. 41 (1995): 55–68.

Andermann, Jens. *The Optic of the State: Visuality and Power in Argentina and Brazil.* University of Pittsburgh Press, 2007.

Andermann, Jens. *Tierras en trance: Arte y naturaleza después del paisaje.* Metales Pesados, 2018.

Bernucci, Leopoldo. *A imitação dos sentidos: Prógonos, contemporâneos e epígonos de Euclides da Cunha.* EDUSP, 1995.

Campos Johnson, Adriana. *Sentencing Canudos: Subalternity in the Backlands of Brazil.* University of Pittsburgh Press, 2010.

Chakrabarty, Dipesh. "The Climate of History: Four Theses." *Critical Inquiry* 35, no. 2 (2009): 197–222.

Chakrabarty, Dipesh. "Anthropocene Time." *History and Theory* 57, no. 1 (2018): 5–32.

Clark, Nigel. "Geo-politics and the Disaster of the Anthropocene." *Sociological Review* 62, no. 1 (2014), 19–37.

Clark, Nigel. "Politics of Strata." *Theory, Culture, and Society* 34, no. 2-3 (2017): 1–21.

Cunha, Euclides da. "Fazedores de desertos." *O estado de São Paulo*, October 22, 1901.

Cunha, Euclides da. *Obra completa.* Nova Aguilar, 1995.

Cunha, Euclides da. *Backlands: The Canudos Campaign.* Translated by Elizabeth Lowe. Penguin, 2010.

Cunha, Euclides da. *Canudos e outros temas.* Senado Federal, 2003.

DeLoughrey, Elizabeth, and George B. Handley. Introduction to *Postcolonial Ecologies: Literatures of the Environment.* Edited by Elizabeth DeLoughrey and George B. Handley, 2–39. Oxford University Press, 2011.

Fonseca, Carlos. *The Literature of Catastrophe: Nature, Disaster, and Revolution in Latin America.* Bloomsbury Academic, 2020.

Galvão, Walnice Nogueira. *No calor da hora: A Guerra de Canudos nos jornais.* Ática, 1977.

Galvão, Walnice Nogueira. *Saco de gatos: Ensaios críticos.* Duas Cidades, 1976.

Guha, Ramachandra. "Radical American Environmentalism and Wilderness Preservation: A Third World Critique." *Environmental Ethics* 11, no. 1 (1989): 71–83.

Huggan, Graham and Helen Tiffin. Introduction to *Postcolonial Ecocriticism: Literature, Animals, Environment.* Edited by Graham Huggan and Helen Tiffin, 1–24. Routledge: 2010.

Kittler, Friedrich A. *Discourse Networks, 1800/1900.* Translated by Michael Metteer with Chris Cullens. Stanford University Press, 1990.

Levine, Robert. *Vale of Tears: Revisiting the Canudos Massacre in Northeastern Brazil, 1893–1897.* University of California Press, 1992.

Nielson, Rex P. "Nefarious Agents of Ecological Collapse: 'The Desert Makers' by Euclides da Cunha." *ISLE: Interdisciplinary Studies in Literature and Environment* 28, no. 1 (2021): 291–311.

Nielson, Rex P. "Ecological Thought in Euclides da Cunha's Correspondence and Writings about the Amazon." *Hispanic Issues On Line* 16 (2014): 17–34.

Pratt, Mary Louise. *Imperial Eyes: Travel Writing and Transculturation*. Routledge, 1992.

Price, Rachel. *The Object of the Atlantic: Concrete Aesthetics in Cuba, Brazil, and Spain, 1868–1968*. Northwestern University Press, 2014.

Souza, Ronaldes de Melo e. *A geopoética da Euclides da Cunha*. EdUERJ, 2009.

Rudwick, Martin J. S. *Earth's Deep History: How It Was Discovered and Why It Matters*. University of Chicago Press, 2014.

Tosta, Luciano. "O 'sublime' sertão em *O sertanejo, Os sertões* e *Grande sertão: Veredas*." *Letras de hoje* 36, no. 1 (2001): 7–26.

Uriarte, Javier. *The Desertmakers: Travel, War, and the State in Latin America*. Routledge, 2020.

Zalasiewicz, Jan. "The Extraordinary Strata of the Anthropocene." In *Environmental Humanities: Voices from the Anthropocene*. Edited by Serpil Oppermann and Serenella Iovino, 115–32. Rowman & Littlefield, 2017.

Acknowledgments

This volume began as a conference panel sponsored by the Nineteenth-Century Latin American section of the Modern Language Association. We are grateful to the MLA and to section members Bram Acosta, Mayra Bottaro, Nathalia Bouzaglo, and Sebastián Díaz-Duhalde.

Innovative interdisciplinary work of this kind depends on the collaboration of archivists, curators, librarians, and other professionals. We are deeply grateful for the individuals who have helped us secure images and texts and the institutions that have chosen to dedicate valuable resources to historical collections in the humanities and natural sciences.

The authors would like to thank in particular the Colección Patricia Phelps de Cisneros; the Macaulay Library at the Cornell Lab of Ornithology; the Biblioteca Nacional de Chile; the Lenhardt Library of the Chicago Botanic Garden; the Soledad Acosta de Samper Digital Library at the Biblioteca Nacional de Colombia; Mexico's Instituto Nacional de las Bellas Artes y Literatura; the New York Public Library; Siemens Historical Institute; the Museo de Barro, Asunción; the Fundación Rafael Barrett; and the Museo de Arte Contemporáneo Atchugarry.

We would also like to express our deep appreciation to the Colección Viviane y Enrique Manhard for permission to reproduce Juan Manuel Blanes's painting Gaucho en el campo on the cover of our book.

We are grateful to Andrea Rosenberg for her translation of Gisela Heffes's chapter and to Omal Jamal for preparing the index.

Lesley Wylie would like to thank the School of Arts, Media, and Communication at the University of Leicester. Jennifer L. French would like to thank the Dean of Faculty at Williams College and the Estate of Robert F. Rosenburg.

Finally, this book would not have come to fruition without the exceptional vision and skill of our colleagues at Vanderbilt University Press.

The authors would like to thank Gianna Mosser, Joell Smith-Borne, Zachary Gresham, Steven Rodriguez, Patrick Samuel, Jenna Phillips, and Alissa Faden.

Contributors

JENS ANDERMANN is a professor of Spanish and Portuguese at NYU and an editor of the *Journal of Latin American Cultural Studies*. He is the author of *Jardín* (2023), *Entranced Earth: Art, Extractivism, and the End of Landscape* (2023; Spanish edition 2018), *New Argentine Cinema* (2011), *The Optic of the State* (2007), and *Mapas de poder* (2000). He has co-edited, among other volumes, *Handbook of Latin American Environmental Aesthetics* (2023), *Natura: Environmental Aesthetics After Landscape* (2018), and—with the *Journal*'s editorial collective— *Latin American Cultural Studies: A Reader* (2018).

RONALD BRIGGS studies the convergence of education and literary theory and is a professor in the Department of Spanish and Latin American Cultures at Barnard College. His publications include *Tropes of Enlightenment in the Age of Bolívar: Simón Rodríguez and the American Essay at Revolution* (2010) and *The Moral Electricity of Print: Transatlantic Education and the Lima Women's Circuit, 1876–1910* (2017), which was awarded the Best Book Prize by the Nineteenth-Century Section of LASA in 2018. He is co-editor of volume 2 of *Latin American Literature in Transition* (2023).

JENNIFER L. FRENCH is the Rosenburg Professor of Environmental Studies and Spanish at Williams College. She is co-editor (with Gisela Heffes) of *The Latin American Ecocultural Reader* (2020) and co-editor (with Felipe Martínez-Pinzón) of *La vorágine: Centenario de un clásico latinoamericano* (2024). She is the author of *Nature, Neocolonialism and the Spanish American Regional Writers*.

GISELA HEFFES is a writer and a professor of Latin American Literature and Culture at Johns Hopkins University. Her most recent publications are the co-edited volumes *The Latin American Ecocultural Reader* (2020), *Pushing Past the Human in Latin American Cinema* (2021), *Un gabinete para el futuro* (2022), and *Turbar la quietud* (2023). She is the author of *Visualizing Loss in Latin America: Biopolitics, Waste, and the Urban Environment* (2023). With George Handley she was the 2022–2024 co-president of ASLE (The Association for the Study of Literature and Environment).

AARTI MADAN is an associate professor of Spanish, director of the Buenos Aires Project Center, and co-director of Latin American & Caribbean Studies at Worcester Polytechnic Institute. In 2013, she was awarded the Romeo Moruzzi Young Faculty Award for Innovation in Undergraduate Education, and, in 2020, she formed part of a grant team funded by the US Department of Education to enhance WPI's STEM curriculum with Latin American and Caribbean studies. Aarti specializes in literary and visual culture from the nineteenth century to present day, focusing primarily on the spatial humanities and the aesthetics of politics. While her first book, *Lines of Geography in Latin American Narrative* (2017), takes a geo- and ecocritical approach to examine nation-building in nineteenth-century Argentina and Brazil, her current project examines Argentine cultural production through a transpacific lens and is provisionally titled *Sensing the Subcontinent: India in the Argentine Cultural Imaginary.*

VANESA MISERES specializes in nineteenth- and early twentieth-century Latin American literature, with an emphasis on South America. Her areas of research are travel writing, war literature, women writers, gender, cultural, and food studies. She is the author of *Mujeres en tránsito: Viaje, identidad y escritura en Sudamérica* (1830–1910) and co-editor of *Food Studies in Latin America: Perspectives on the Gastronarrative.* Miseres is finishing a book on women's war writing and feminist ideas in Latin America and preparing a new manuscript on the cultural history of gender and science in Latin America, covering the time period of 1850 to 1950.

LIZABETH PARAVISINI-GEBERT is a professor in the Department of Hispanic Studies and the Environmental Studies Program at Vassar College, where she holds the Sarah Tod Fitz Randolph Distinguished Professor Chair. She is the author of a number of books, among them *Phyllis Shand Allfrey: A Caribbean Life* (1996), *Jamaica Kincaid: A Critical Companion* (1999), *Creole Religions of the Caribbean* (with Margarite Fernández Olmos, 2003), *Literatures of the Caribbean* (2008), and the forthcoming *Extinctions: The Eco-*

logical Cost of Colonization in the Caribbean. She co-edits *Repeating Islands*, a blog on Caribbean culture, with Ivette Romero-Cesareo.

JORGE QUINTANA NAVARRETE is an assistant professor of Spanish at Dartmouth College. His research interests include Mexican culture (nineteenth and twentieth centuries), utopian studies, environmental humanities, and critical theory. He has published chapters in edited volumes and peer-reviewed articles in *Hispanic Review, Journal of Latin American Cultural Studies, Revista Hispánica Moderna*, and *Revista Canadiense de Estudios Hispánicos*, among others. He is the author of *Biocosmism: Vitality and the Utopian Imagination in Postrevolutionary Mexico*.

CATALINA RODRÍGUEZ is an assistant professor in the Department of Romance Studies at Boston University. Her research focuses on the areas of women's literature, pseudonymous writing, gender and sexuality studies, nineteenth-century Latin American cultural studies, and queer literature. Her publications have appeared in the *Latin American Literary Review* and the *Revista Taller de letras*.

EMMANUEL A. VELAYOS LARRABURE is an assistant professor of Latin American and Caribbean studies at CUNY's Hostos CC. He also teaches global history at the Cooper Union for the Advancement of Science and Art. His research explores the relationship between lettered and performative cultures in Latin America, through the lenses of race studies, media aesthetics, and postcolonialism. He has published articles about race, media experiments, and the embodied dimension of intellectual practices in several peer-reviewed journals, including *Hispanic Review, Chasqui, Revista de Crítica Literaria Latinoamericana, Revista Hispánica Moderna, Decimonónica*, and *A Contracorriente*. His article "Painting Words, Drawing Republics: Embodied Arts and New Beginnings in Simón Rodríguez" (*Hispanic Review*) received the 2020 LASA Prize for Best Article of the Nineteenth-Century Studies Section. He has also received grants from the Mellon Foundation, the National Endowment for the Humanities, L'École des Hautes Études en Sciences Sociales, NYU, Hostos CC, and PSC-CUNY.

LESLEY WYLIE is a professor of Latin American studies at the University of Leicester, UK. She is the author of the books *Colonial Tropes and Postcolonial Tricks: Rewriting the Tropics in the novela de la selva* (2009);

Colombia's Forgotten Frontier: A Literary Geography of the Putumayo (2013); and *The Poetics of Plants in Spanish American Literature* (2020). She is particularly interested in the intersections between literature and the environment, with a focus on the Amazon, Cuba, and more recently Argentina. She is currently working on a book project about the Anglo-Argentine author and naturalist W. H. Hudson. She is also senior associate editor of the *Bulletin of Spanish Studies*.

Index

Page numbers in *italic refer to figures.*

www.ingramcontent.com/pod-product-compliance
Lightning Source LLC
Chambersburg PA
CBHW060612030726
47498CB00005B/1649